Tucholsky Wagner Zola Scott Sydow Freud Schlegel
Turgenev Wallace Fonatne
Twain Walther von der Vogelweide Fouqué Friedrich II. von Preußen
Weber Freiligrath Frey
Fechner Weiße Rose von Fallersleben Kant Ernst Frommel
Fichte Richthofen
Hölderlin
Engels Fielding Eichendorff Tacitus Dumas
Fehrs Faber Flaubert Eliasberg Ebner Eschenbach
Feuerbach Maximilian I. von Habsburg Fock Eliot Zweig Vergil
Ewald
Goethe Elisabeth von Österreich London
Mendelssohn Balzac Shakespeare Dostojewski Ganghofer
Lichtenberg Rathenau Doyle Gjellerup
Trackl Stevenson Hambruch
Mommsen Tolstoi Lenz Hanrieder Droste-Hülshoff
Thoma von Arnim Hägele Hauff Humboldt
Dach Verne
Reuter Rousseau Hagen Hauptmann Gautier
Karrillon Garschin Baudelaire
Defoe Hebbel
Damaschke Descartes Hegel Kussmaul Herder
Wolfram von Eschenbach Dickens Schopenhauer Rilke George
Bronner Darwin Melville Grimm Jerome Bebel
Campe Horváth Aristoteles Federer Proust
Bismarck Vigny Barlach Voltaire Herodot
Gengenbach Heine
Storm Casanova Tersteegen Gilm Grillparzer Georgy
Chamberlain Lessing Langbein
Brentano Lafontaine Gryphius
Strachwitz Claudius Schiller Kralik Iffland Sokrates
Bellamy Schilling
Katharina II. von Rußland Gerstäcker Raabe Gibbon Tschechow
Löns Hesse Hoffmann Gogol Wilde Gleim Vulpius
Luther Heym Hofmannsthal Morgenstern
Roth Klee Hölty Goedicke
Heyse Klopstock Kleist
Luxemburg Puschkin Homer Mörike Musil
Machiavelli La Roche Horaz
Navarra Aurel Musset Kierkegaard Kraft Kraus
Lamprecht Kind Moltke
Nestroy Marie de France Kirchhoff Hugo
Laotse Ipsen Liebknecht
Nietzsche Nansen Ringelnatz
Marx Lassalle Gorki Klett Leibniz
von Ossietzky May Lawrence Irving
vom Stein
Petalozzi Knigge
Platon Pückler Michelangelo Kafka
Sachs Poe Kock
de Sade Praetorius Mistral Liebermann Korolenko
Zetkin

The publishing house tredition has created the series **TREDITION CLASSICS**. It contains classical literature works from over two thousand years. Most of these titles have been out of print and off the bookstore shelves for decades.

The book series is intended to preserve the cultural legacy and to promote the timeless works of classical literature. As a reader of a **TREDITION CLASSICS** book, the reader supports the mission to save many of the amazing works of world literature from oblivion.

The symbol of **TREDITION CLASSICS** is Johannes Gutenberg (1400 – 1468), the inventor of movable type printing.

With the series, tredition intends to make thousands of international literature classics available in printed format again – worldwide.

All books are available at book retailers worldwide in paperback and in hardcover. For more information please visit: www.tredition.com

tredition was established in 2006 by Sandra Latusseck and Soenke Schulz. Based in Hamburg, Germany, tredition offers publishing solutions to authors and publishing houses, combined with worldwide distribution of printed and digital book content. tredition is uniquely positioned to enable authors and publishing houses to create books on their own terms and without conventional manufacturing risks.

For more information please visit: www.tredition.com

How Two Boys Made Their Own Electrical Apparatus Containing Complete Directions for Making All Kinds of Simple Apparatus for the Study of Elementary Electricity

Thomas M. (Thomas Matthew) St. John

Imprint

This book is part of the TREDITION CLASSICS series.

Author: Thomas M. (Thomas Matthew) St. John
Cover design: toepferschumann, Berlin (Germany)

Publisher: tredition GmbH, Hamburg (Germany)
ISBN: 978-3-8491-7181-0

www.tredition.com
www.tredition.de

Copyright:
The content of this book is sourced from the public domain.

The intention of the TREDITION CLASSICS series is to make world literature in the public domain available in printed format. Literary enthusiasts and organizations worldwide have scanned and digitally edited the original texts. tredition has subsequently formatted and redesigned the content into a modern reading layout. Therefore, we cannot guarantee the exact reproduction of the original format of a particular historic edition. Please also note that no modifications have been made to the spelling, therefore it may differ from the orthography used today.

TABLE OF CONTENTS.

Chapter.

I.	Cells and Batteries,
II.	Battery Fluids and Solutions,
III.	Miscellaneous Apparatus and Methods of Construction,
IV.	Switches and Cut-Outs,
V.	Binding-Posts and Connectors,
VI.	Permanent Magnets,
VII.	Magnetic Needles and Compasses,
VIII.	Yokes and Armatures,
IX.	Electro-Magnets,
X.	Wire-Winding Apparatus,
XI.	Induction Coils and Their Attachments,
XII.	Contact Breakers and Current Interrupters,
XIII.	Current Detectors and Galvanometers,
XIV.	Telegraph Keys and Sounders,
XV.	Electric Bells and Buzzers,
XVI.	Commutators and Current Reversers,
XVII.	Resistance Coils,
XVIII.	Apparatus for Static Electricity,
XIX.	Electric Motors,
XX.	Odds and Ends,
XXI.	Tools and Materials,

A WORD TO BOYS.

The author is well aware that the average boy has but few tools, and he has kept this fact constantly in mind. It is a very easy matter for a skilled mechanic to make, with proper tools, very fine-looking pieces of apparatus. It is *not* easy to make good apparatus with few tools and a limited amount of skill, *unless* you follow *simple methods*.

By following the methods given, any boy of average ability can make the apparatus herein described.

Most of the illustrations have been made directly from apparatus constructed by young boys.

It is impossible to describe the different pieces of apparatus in any special or logical order. It is taken for granted that you have some book of simple experiments and explanations to serve as a guide for the order, and to give you an idea of just the apparatus needed for the special experiments.

It would be foolish to start in and make all the apparatus described, without being able to intelligently use it in your experiments. *Take up a systematic course of simple experiments, and make your own apparatus, as needed.*

Before making any particular piece of apparatus, read what is said about the other pieces of the same general nature. This will often be a great help, and it may suggest improvements that you would like to have.

In case your apparatus does not work as expected, read the directions again, and see if you have followed them. Wrong connections, poor connections, short circuits, broken wire, etc., will make trouble. With a little patience and care you will be able to locate and correct any troubles that may come up in such simple apparatus.

Thomas M. St. John.

How Two Boys Made Their Own Electrical Apparatus

CHAPTER I.

CELLS AND BATTERIES.

APPARATUS 1.

1. Carbon-Zinc Cell. Fig. 1. If you have some rubber bands you can quickly make a cell out of rods of zinc and carbon. The rods are kept apart by putting a band, B, around each end of both rods. The bare wires are pinched under the upper bands. The whole is then bound together by means of the bands, A, and placed in a tumbler of fluid, as given in App. 15. This method does not make first-class connections between the wire and rods. (See § 3.)

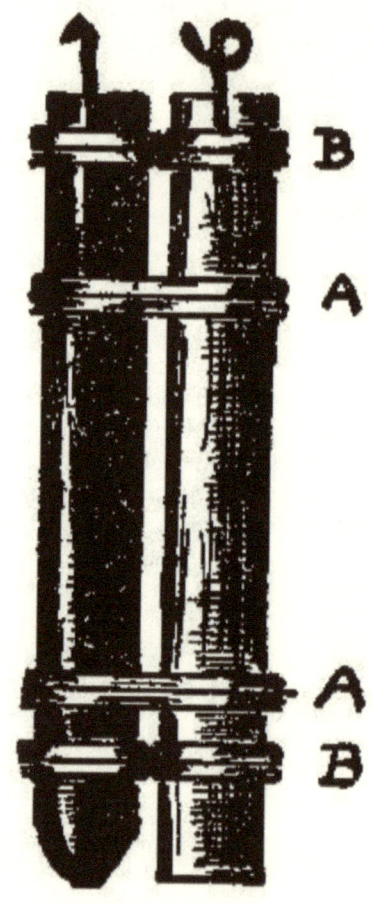

Fig. 1.

APPARATUS 2.

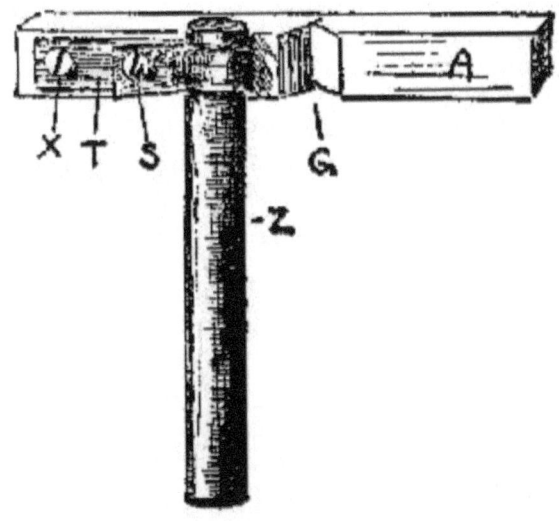

Fig. 2.

2. Carbon-Zinc Cell. Fig. 2. In case you want to make your cell out of carbon and zinc rods, and do not have any means of making holes for them in the wood, as in App. 3 and 4, you will find this method useful. Cut grooves, G, into one side of the wood, A, which should be about 4½ × 1 × ½ in. The grooves should be quite deep, and so placed that the rods will be about ¼ in. apart. A strip of tin, T, ½ in. wide, should be bent around each [Pg 6] rod. The screw, S, put through the two thicknesses of tin will hold the rod in place. Another screw, X, acts as a binding-post. The zinc rod only is shown in Fig. 2. The carbon rod is arranged in the same way. Use the fluid of App. 15.

3. Note. When the bichromate solution of App. 15 is used for cells, the strong current is given, among other reasons, because the zinc is rapidly eaten up. This action goes on even when the circuit is broken, so always remove and wash the zinc as soon as you have finished.

APPARATUS 3.

4. Carbon-Zinc Cell. Fig. 3. The wooden cross-piece, A, is 4½ × 1 × ½ in. The carbon and zinc rods, C and Z, are 4 in. long × ½ in. in diameter. The holes are bored, if you have a brace and bit, so that they are ¾ in. apart, center to center. This makes the rods ¼ in. apart. To make connections between the rods and outside wires, cut a shallow slot at the front side of each hole, so that you can put a narrow strip of tin or copper, B, in the hole by the side of each rod. Setscrews, S, screwed in the side of A, will hold the rods in place, and at the same time press the strips, B, against them. Connections can easily be made between wire and B by using a spring binding-post, D, or by fastening the wire direct to the strips, as shown in App. 4.

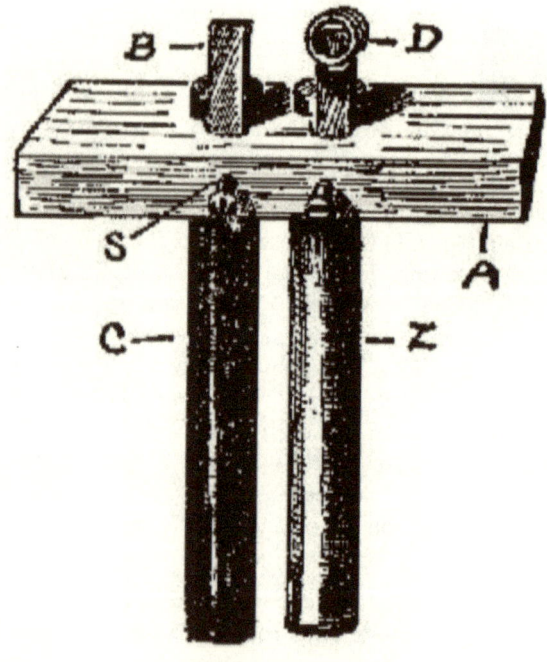

Fig. 3.

Use the battery fluid given in App. 15, and use a tumbler for the battery jar. This cell will run small, well-made motors, induction coils, etc. (See § 3.)

[Pg 7]

APPARATUS 4.

5. *Carbon-Zinc Cell.* Fig. 4. The general construction of this cell is the same as that of App. 3. There are 2 carbons, C, each $4 \times \frac{1}{2}$ in. The holes for these are bored in A 1¼ in. apart, center to center. The zinc rod, Z, is a regular battery zinc, $6 \times \frac{3}{8}$ in., and has a binding-post, Y, of its own. The rods, C, are held in A, and connections are made as explained in App. 3.

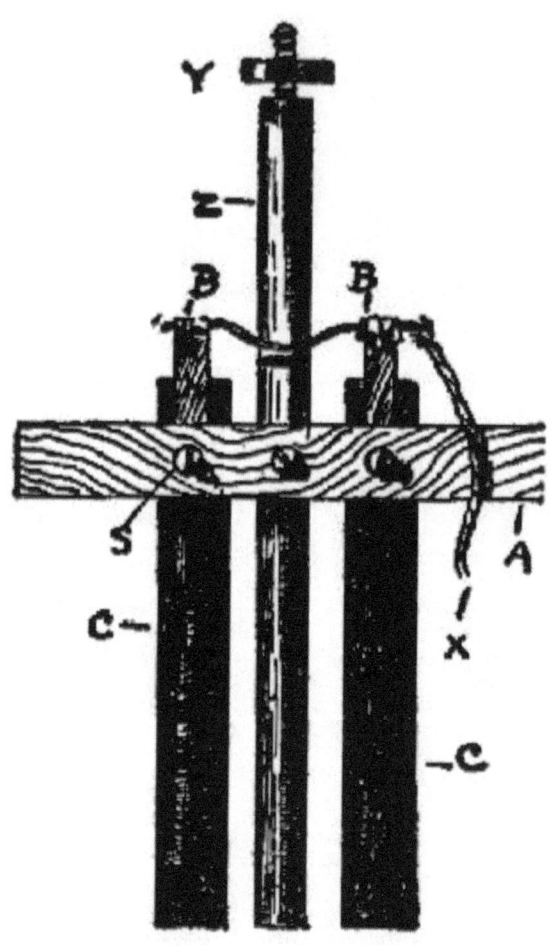

Fig. 4.

The wire, X, is fastened direct to the strips, B, as shown. When ready to use this cell, be sure that the wire connecting the carbons does not touch Z. (Why?) The other wire is connected to Y. The wooden piece is 4½ × 1 × ½ in. Use the battery fluid of App. 15 in a tumbler. This cell will run small motors, and is good for induction coils, etc. (See § 3.)

APPARATUS 5.

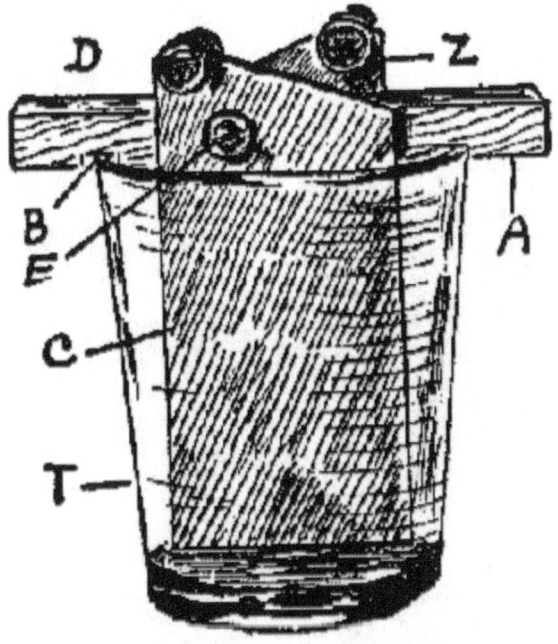

Fig. 5.

6. *Experimental Cell.* Fig. 5. Cut a strip each of copper, C, and zinc, Z. (See list of materials.) They should be about 2 in. wide and 4 in. long. Punch a hole through each, one side of the center, for screws, E. The wooden cross-piece, A, should be 4½ × 1 × ⅞ in. The battery-plates, or elements, should be screwed to this, taking care that the screws, E, do not touch each other. If the holes are [Pg 8] made in the position shown in Fig. 5, the screws can be arranged some distance apart.

The wires leading from the cell may be fastened under the screws with copper burs, or spring binding-posts (App. 42) can be slipped on the top of the plates.

The solution to be used will depend upon what the cell is to do. For simple experiments use the dilute acid (App. 14). If for small motors, use the formula given in App. 15. The zinc should be well amalgamated. (App. 20.)

APPARATUS 6.

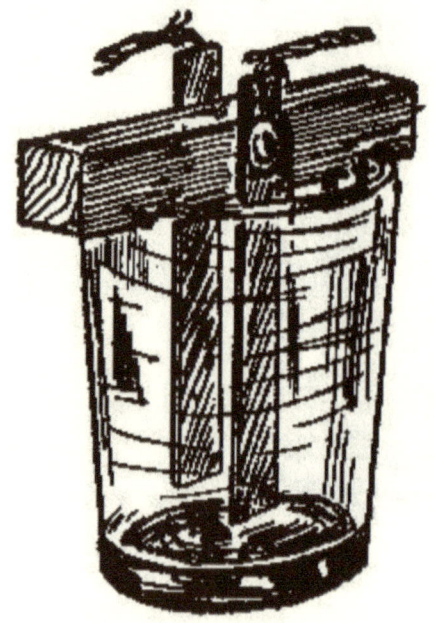

Fig. 6.

7. *Experimental Cell.* Fig. 6. In some experiments a comparison is made between cells with large plates and cells with small ones. This form will be convenient to use where narrow plates are desired. Those shown are 4 × ½ in. They are screwed to the cross-piece, which is 4½ × 1 × ⅞ in. Do not let the screws touch each other. The wires are fastened under the screw-heads.

APPARATUS 7.

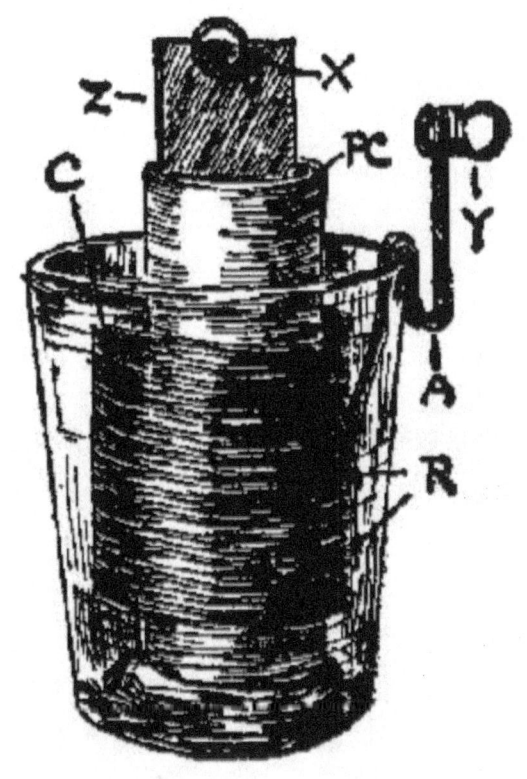

Fig. 7.

8. Experimental Two-fluid Cell. Fig. 7. This cell has a zinc strip, Z, and copper cylinder, C, for the "elements." The porous cup, P C, is fully described in App. 11. Z is 5 × 1 in., and should be well amalgamated [Pg 9] (App. 20). (Study reasons for amalgamation.) A zinc rod, like that shown in Fig. 4, may be used instead of the strip. The copper cylinder, C, nearly surrounds P C, and is made from a piece of thin sheet-copper, 6 × 2 in. The narrow strip, or leader, A, is 5 × ½ in. To fasten it to C, punch two small holes in C and A, put short lengths of stout copper wire through the holes, and hammer them down so that they will act as rivets, R. C can be hung centrally in the tumbler by bending A as shown. Y and X are spring binding-posts

(App. 42). The battery wires can be fastened directly to Z and A, as suggested in Fig. 4.

9. Setting up the Cell. Arrange as in Fig. 7, but remove Z from P C. Pour some of the acid solution of App. 14 into P C until it stands about 2½ in. deep, and at once pour the copper solution of App. 16 in the tumbler, on the outside of P C, until it stands at the same height as the liquid in P C. As soon as the liquids have soaked into P C, you can put Z in place, when the cell will be ready for use. Remove and wash Z, when you have finished, and if you wish to use this cell occasionally, remove the liquids and wash P C thoroughly in water. When dry it will be as good as new. The acid rapidly acts upon Z, so it is better to remove Z if you wish to leave the experimenting even for a few minutes only.

Put a few crystals of copper sulphate (blue vitriol) in the tumbler under the copper, to keep the copper solution saturated. (See textbook for the chemical action in this two-fluid cell.)

APPARATUS 8.

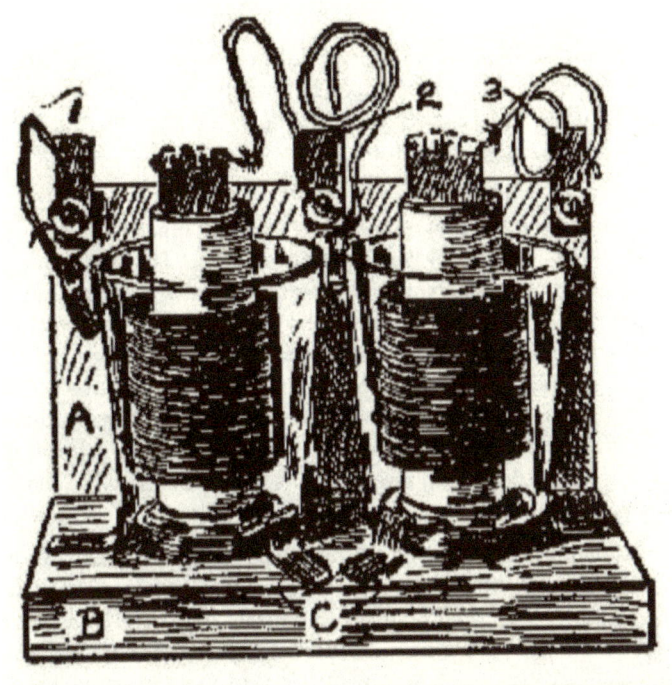

Fig. 8.

10. Two-fluid Battery. Fig. 8. When two or more cells are joined together the combination is called a battery. Fig. 8 shows two experimental cells joined in series. (Study methods of joining cells.) For convenience, [Pg 10] and to keep them from being easily overturned, a frame has been made for them. The base, B, is $8 \times 4 \times 7/8$ in. To the back of this is nailed the upright board, A, $8 \times 4\frac{1}{2} \times \frac{1}{2}$ in. On the top of A are 3 binding-posts, 1, 2, 3, which consist of metal strips $1\frac{1}{4} \times \frac{1}{2}$ in. At the lower ends are screws which are connected with the cells, as shown. Spring binders can be easily slipped on and off the upper ends of the strips, so that one or two cells can be used at will. Bent strips, C, are nailed to B, to hold the tumblers firmly in place. This framework is not necessary, of course, to the proper working of the battery, but with it you are much less liable to upset the cells.

APPARATUS 9.

11. *Gravity Cell.* Fig. 9. In the two-fluid cell of App. 7 the fluids were kept apart by the porous cup. The gravity cell is really a two-fluid cell in which the two liquids are kept separate by the joint action of the current and the force of gravity. This cell is used for telegraph lines and for other closed-circuit work.

12. Construction. The zinc and copper, Z and C, Fig. 9, can be purchased about as cheaply as you can make them. There are many forms of the zincs, the one shown being called the crow-foot shape. The copper may be star-shaped, or as shown. If you wish to make C, use thin sheet-copper. Brush copper, 1¾ in. wide, is excellent for the purpose. Use a piece 12 or 15 in. long, and fasten to one end of it a copper wire, W, which must be [Pg 11] covered with paraffined paper, or with rubber or glass tubing, where it passes up through the zinc sulphate solution and near Z. The glass jar, J, may be made from a large glass bottle. (See index for battery jars.)

13. To Set Up the Cell. (A) Place C upon the bottom of J, with W in the position shown. (B) Put in enough copper sulphate crystals to cover the bottom of J, but do not try to entirely cover C. At the start ½ lb. will be enough. (C) Pour in clean water until J is half full. (D) In another vessel dissolve 1 or 2 oz. of zinc sulphate in enough water to complete filling, J. (E) Hang Z in place (Fig. 9). Z must never touch C. They should be about 3 in. apart. A wire is attached to Z by the screw, S, and the hole, H. (F) Pour the zinc sulphate solution into J until it is within an inch of the top. It should cover Z.

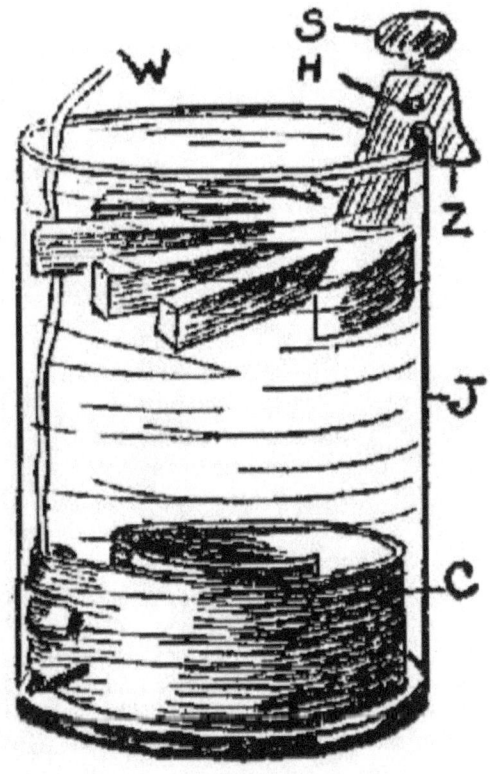

Fig. 9.

(G) Connect the wires leading from Z and C to your sounder and key. (See diagram.) The cell will be weak at first, and it may not be able to run your sounder. If this is the case, "short-circuit" it by allowing the current to run around and around through the sounder and key, the switch being closed. You may also "short-circuit" the cell by joining the two wires together. This will, in a few hours, make the dividing line between the blue and white quite distinct, when the cell will be stronger. If you have a short line only, the battery may be short-circuited through your sounder or other coils of wire for 5 or 6 hours a day, without working it too much. It may be necessary to draw off some of the clear zinc sulphate, [Pg 12] replacing it with clear water, if the blue line gets too low. Add water occasionally to make up for evaporation.

14. Regulating. The two solutions are kept apart by gravity, as the copper sulphate is heavier than the zinc sulphate. The dividing line between the blue and white solutions is fairly clear when the battery works well, and it should be about half way between C and Z, or about at J, Fig. 9. Never allow the blue to get as high as Z, as this indicates that the cell is not worked enough. The dividing line can be lowered by allowing it to run a buzzer or bell for a few hours, or by simply short-circuiting it. If the blue gets much below J it indicates that you are working the cell too hard, or that you need more copper sulphate. The harder the cell works, the more zinc sulphate is formed, and the lower the dividing line becomes.

15. Gravity Batteries of two more cells are needed when used on telegraph lines. You will need 1 cell to each sounder; that is, for a short line in the house with two sounders, use 2 cells. If you use a few hundred feet of wire running to a friend's house, use 3 cells. They must be joined in series; that is, the copper of one to the zinc of the other. (See diagram of complete telegraph line.) Do not use ground connections for short lines and home-made sounders; use a return wire. Do not use different kinds of cells upon the same line.

APPARATUS 10.

16. Storage Battery. To show the principle of storage batteries it is only necessary to use two plates of lead dipped in the battery fluid of App. 14. The cell may be made as in App. 5, Fig. 5, the only difference being that both plates are of sheet-lead. It will be an advantage to make the plates rough by hammering against [Pg 13] them a coarse file. (See explanations and experiments with this form of cell in text-book.)

APPARATUS 11.

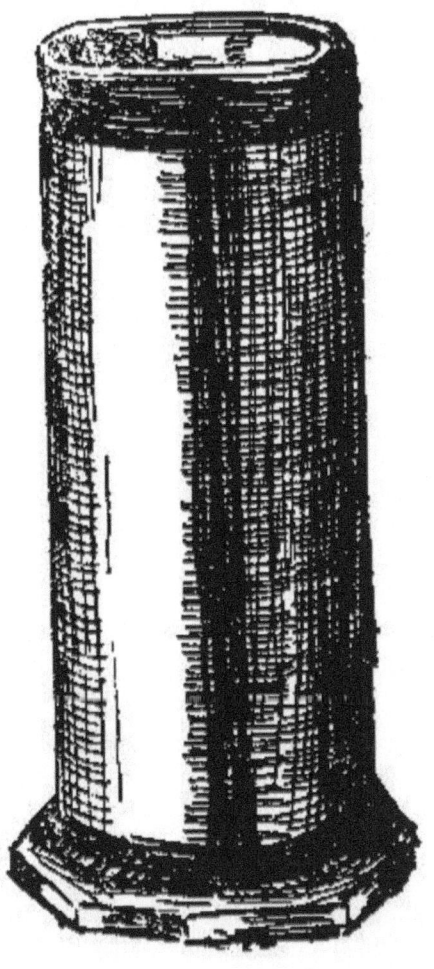

Fig. 10.

17. *Porous Cups for Two-fluid Cells.* Fig. 10. Very good porous cups can be made from ordinary blotting-papers, the average ones measuring 9½ × 4 in. White ones should be used, so that you will not be bothered with the color coming out. Soak the edge along one end of the blotter in paraffine (Index) for about ¼ in. When this is cold, roll the blotter into the form of a cylinder that is a little over 1

23

in. inside diameter, and have the paraffined end on the outside. This will make 2 thicknesses of paper all around, and a little to spare. Rub a hot nail over the paraffine to melt it, and stick the end to the cylinder. By putting on a little more paraffine along the edge where the end laps over, a good solid cylinder can be made. The cylinder should be strengthened still more by dipping each end into melted paraffine for about ⅛ in. The dark stripes around the ends and down the front of the cylinder (Fig. 10) are to represent the paraffine. Cut out a bottom about ¼ in. larger all around than the cylinder. This may be paraffined to make it stiff. It should be fastened to the cylinder with paraffine. Paraffine is not acted upon or softened by water or acid, as is the case with glue.

APPARATUS 12.

18. Porous Cups for Two-fluid Cells. Instead of the blotters of App. 11, you can use short lengths of mailing-tubes, which are used to protect pictures, etc., when sent by mail. If you find that the particular tube [Pg 14] tends to unwind when soaked, you can use a little paraffine along the edges of the spiral, as suggested in App. 11. Bottoms can be made for the cups as before.

APPARATUS 13.

19. Porous Cups for Two-fluid Cells. Ordinary unglazed earthen flower-pots make good cups. The hole in the bottom should be closed with a cork, or by fastening a piece of pasteboard over the hole with paraffine. The pasteboard may be fastened to the under side of the bottom more easily than to the upper side.

20. Note. It is a good idea to soak the top edge of porous cups for about ¼ in. in paraffine to keep the solutions from crawling up by capillary attraction. If the solutions constantly evaporate from the soaked tops of the cups, they not only waste but they get the whole thing covered with crystals.

[Pg 15]

CHAPTER II.

BATTERY FLUIDS AND SOLUTIONS.

21. Sulphuric Acid. This acid must be handled with great care, as it (the concentrated) is very strong, and will burn the hands, eat holes in clothing, carpets, etc.; it will even char wood. Do not let any of it drop anywhere accidentally. If you wish to pour concentrated acid into a bottle, place the bottle to be filled upon a plate, and wipe all drops of acid from the outside of it afterward. The concentrated acid should be kept in tightly-corked bottles, as it absorbs moisture from the air very rapidly. Ordinary corks should be paraffined if they are to be used in acid bottles, or they will be soon eaten up.

22. *Mixing.* When sulphuric acid and water are mixed, considerable heat is produced. *Never pour water into the acid*, as the heat would be produced so rapidly that the vessel containing the mixture might break. *Always* pour the acid into the water, and thoroughly stir the mixture at the same time. Earthen vessels do not break when heated as easily as glass ones. The mixing may be done in ordinary glass fruit-jars, if care be taken to pour the acid *slowly* into the water. The jars should be set in some larger dish, or in the sink, before adding the acid. If they get too hot, allow them to cool a little before proceeding with the mixing. As the acid is much heavier than water, it will immediately sink to the bottom of the jar, unless constantly stirred.

23. There are different grades of acid upon the market. For battery purposes you do not need the chemically pure [Pg 16] (C P) acid. The ordinary "commercial acid" is all right, even though it is a little dark in color. You can get this at any drug-store. Get 5 or 10 cents' worth at a time.

APPARATUS 14.

24. *Battery Fluid for Simple Cells.* For the simple cell (App. 5), when it is to be used for experiments with detectors or in the study of polarization, etc., a very dilute acid is best. Mix 1 fluid ounce of commercial acid with 1 pint of water. This will make 17 fluid ounces (See App. 19), and your mixture will be one-seventeenth acid. Make up a pint or quart bottle of this at a time, and label it with the date:

> Dilute sulphuric acid.
> 1 part acid, 16 parts water.
> Apparatus 14.

25. *Note.* Do not fail to paste a label on all bottles as soon as you have put anything into them. Give the date, contents, and any other information that will help you to reproduce the mixture again. Do not write on them any abbreviations or other things that you will soon forget.

APPARATUS 15.

26. *Battery Fluid; Bichromate Solution.* For running small motors, shocking coils, etc., this solution will be found good when used with the zinc and carbon elements given in App. 3 and 4. The bichromate destroys the hydrogen bubbles which help to polarize cells so rapidly when the plain dilute acid (App. 14) is used. (Study polarization.) The zinc used in this fluid must be well amalgamated (App. 20).

Directions. With 1 quart of cold water placed in a glass or earthen dish, slowly mix 4 fluid ounces of commercial [Pg 17] sulphuric acid. *Read § 22 carefully.* When this gets about cold, add 4 ounces of bichromate of potash. Powdered bichromate will dissolve more quickly than the lump. Keep this fluid in corked bottles, labelled, with date:

Bichromate Battery Fluid.
Apparatus 15.

27. Always take the zinc from this fluid as soon as you have finished experimenting, or even if you have no use for the cell for a few minutes. The zinc and fluid are rapidly destroyed in bichromate cells even when the circuit is open. Always wash the carbon and zinc as soon as you take them from the fluid.

APPARATUS 16.

28. *Battery Fluid.* For 2-fluid cells (App. 7), a saturated solution of copper sulphate (blue vitriol) is needed. Place some of the crystals in a glass jar, with water, stir them around, and add the sulphate as long as it is dissolved. A few extra crystals should be left in the stock bottle so that the solution will always be saturated.

APPARATUS 17.

29. Vinegar Battery Fluid. For a few of the experiments with detectors, etc., good strong vinegar does well as the exciting fluid. This may be used with the copper and zinc or carbon and zinc elements. The amount of current given with vinegar and App. 4 or 5 is sufficient to show many of the simpler experiments.

APPARATUS 18.

30. Battery Fluid. Strong brine, made by dissolving ordinary salt in water, will produce quite a little current with App. 4 or 5. The presence of the current is easily shown with the astatic detectors.

APPARATUS 19.

31. Measures for Water, Acids, etc. If you do not own a graduated glass, such as druggists use for measuring liquids, the following plan will be found useful. In the mixing of battery fluids, etc., while it is not necessary to be absolutely exact, it is necessary to know approximately what you are doing.

An ordinary glass pint fruit jar may be taken as the standard. This holds 16 fluid ounces, or 2 ordinary teacupfuls. A teacupful may then be taken as ½ pint, or 8 fluid ounces. You can probably find a small bottle that will hold 1 or 2 oz., and you can easily tell how much it holds by filling it and counting the number of times it is contained in the pint can.

A slim bottle holding ½ pint can be made into a convenient measuring glass by scratching lines on it with the sharp edge of a hard file. The lines should be placed, of course, so that they will show how much liquid you must put into it to make 1 oz., 2 oz., etc. Instead of the file marks, a narrow strip of paper may be pasted upon the bottle, and the divisions shown by lines drawn upon the paper.

APPARATUS 20.

32. To Amalgamate Battery Plates. To keep the *zinc* plates or rods in cells from being eaten or dissolved when the circuit is opened, they should be amalgamated; that is, they should have a coating of mercury. The local currents (see text-book) aid in rapidly destroying the zinc, unless it is amalgamated. Do not amalgamate copper plates—merely the zinc ones.

33. Place a few drops of mercury in a butter dish. Dip the zinc into the solution of App. 14, then lay it upon a flat board. This is necessary with *thin* sheet-zinc, as it becomes [Pg 19] very brittle when coated with mercury, and will not stand hard rubbing. If you also dip a very narrow piece of tin into the dilute sulphuric acid, you can use this as a spoon and lift one drop of mercury at a time from the butter dish to the zinc. By tapping the tin upon the zinc, the mercury will leave the tin. Put the mercury only where the zinc will be under the solutions in the cell, then rub the drops around with a small cloth that has been dipped in the acid. The zinc will become very bright and silvery, due to the mercury. Do not get too much mercury on it, just enough to give it a thin coat, as it will make the thin zinc so brittle that it will very easily break. Amalgamate both sides of the zinc.

[Pg 20]

CHAPTER III.

MISCELLANEOUS APPARATUS AND METHODS OF CONSTRUCTION.

APPARATUS 21.

34. *For Annealing and Hardening Steel.* (See text-book for reasons why some parts of electrical apparatus should be made of hard steel, while other parts should be made of soft iron.)

35. *To anneal* or soften spring steel so that you can bend it without breaking it, heat it in a candle, gas, or alcohol flame until it is red-hot; allow the steel to cool in the air slowly.

36. *To harden* steel, heat as before, then suddenly plunge the red-hot piece into cold water. This will make the steel very hard and brittle.

Small pieces may be held by pinching them between two pieces of wood. Needles and wires may be stuck in a cork, which will serve as a handle. (See text-book.)

APPARATUS 22.

Fig. 11.

37. Alcohol Lamp. Fig. 11. An alcohol lamp is very useful in many experiments, and it is better than a candle for annealing or hardening steel needles when making small magnets (App. 21). You can make a good lamp by using a small bottle with a wide opening. A vaseline bottle or even an ink bottle will do. Make a hole about ¼ in. in diameter through the cork with a small round file, or burn it through with a hot nail. Make a cylinder of tin about 1½ in. long and just large enough to push through the hole. The tin may be simply rolled up. If [Pg 21] you have glass tubing, use a short length of that instead of the tin. For the wick, roll up some flannel cloth. This should not fit the inside of the tin tube too tightly. The alcohol should be put into the lamp when you want to use it, and that left should be put back into the supply-bottle when you have finished, as alcohol evaporates very rapidly. The flame of this lamp is light-blue in color, and very hot.

Caution. Do not have your supply-bottle of alcohol near the lamp when you light the latter, or near any other flame. The vapor of alcohol is explosive.

APPARATUS 23.

38. *Spool Holder for Wire.* Fig. 12. When winding magnets it is necessary to have the spool of wire so arranged that it will take care of itself and not interfere with the winding. If you have a brace and bit, bore a hole in a base ⅞ in. thick for a ¼ in. dowel. The dowel should fit the hole tight. The spools of wire purchased can then be placed upon the dowel, where they will unwind evenly. The base may be nailed or clamped to a table.

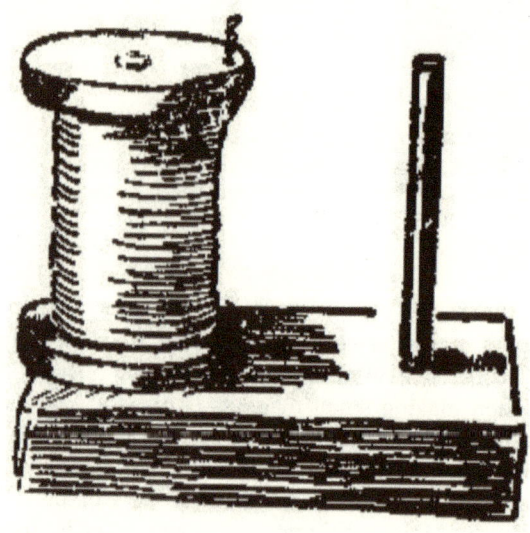

Fig. 12.

APPARATUS 24.

39. *Spool Holder for Wire.* If you have no brace and bit to make App. 23, nail a spool to a wooden base, place a short length of dowel in the spool, and use this combination as a spool holder. Make the dowel fit the spool by winding paper around it.

APPARATUS 25.

40. *To Make Holes in Wood.* If you have a brace and a set of bits, or even a small hand-drill, it will be an easy matter to bore holes in wood. An awl should [Pg 22] be used to make holes for screws, such as those used in making binding-posts, etc., as the wood is very liable to split if a screw is forced into it without a previously-made hole.

Red-hot nails, needles, etc., are easily made to burn holes of desired diameters. They may be heated in a gas flame or by means of the alcohol lamp (App. 22). Flat pieces of hot steel will burn narrow slots, and small, square holes may be made with hot nails.

APPARATUS 26.

Fig. 13.

41. To Make Holes in Sheet-Metal. Fig. 13. Holes may be punched in sheet-tin, copper, zinc, etc., in the following manner: Set a block of hard wood, *W*, on end; that is, place it so that you will pound directly against the *end* of the grain. Lay the metal, *T*, to be punched, upon this, and use a flat-ended punch. A sharp blow upon a good punch with a hammer will make a fairly clean hole; that is, it will cut out a piece of metal, and push it down into the wood. A sharp-pointed punch will merely push the metal aside, and leave a very ragged edge to the hole. A punch may be made of a nail by filing its end flat.

APPARATUS 27.

42. *To Punch Holes through Thick Yokes, etc.* As soon as 5 or 6 layers are to be punched at one operation, the process becomes a little more difficult than that given in App. 26. If you have an anvil, you can place the yoke over one of the round holes in it, and punch the tin right down into the hole, the ragged edges being afterward filed off. Hold the yoke as in App. 79 or 80 for filing. As you will probably have no anvil, lay an [Pg 23] old nut from a bolt upon the end of the block of wood (App. 26), place the metal to be punched over the hole, and imagine that you have an anvil. Very good results may be obtained by this method. The size of nut used will depend upon the size of hole wanted.

APPARATUS 28.

43. *To Straighten Wires.* It is often necessary to have short lengths of wires straight, where they are to be made into bundles, etc. To straighten them, lay one or two at a time upon a perfectly flat surface, place a flat piece of board upon them, then roll them back and forth between the two. The upper board should be pressed down upon the wires while rolling them. If properly done, the wires can be quickly made as straight as needles.

44. *Push-Buttons.* Nearly every house has use for one or more push-buttons. The simple act of pressing your finger upon a movable button, or knob, may ring a bell a mile away, or do some other equally wonderful thing.

APPARATUS 29.

45. *Push-Button.* Fig. 14. This is made quickly, and may be easily fastened to the window or door-casing. One wire is joined to A and the other to C. B is a strip of tin or other metal, about ⅝ in. wide and 2 in. long. It is bent so that it will not touch A unless it is pressed down. This may be placed anywhere, in an electric-bell circuit or other open circuit, where it is desired to let the current pass for a moment only at a time.

Fig. 14.

[Pg 24]

APPARATUS 30.

46. Push-Button. Fig. 15 and Fig. 16. By placing App. 29 in a box, we can make something that looks a little more like a real push-button. Fig. 15 shows a plan with the box-cover removed, and Fig. 16 shows a view of the inside of it, a part of the box being cut away. C, Fig. 15, is a wooden pill-box 1 in. high and 1¾ in. in diameter. Make a ¼ in. hole in the cover of C for the "button," G, which is a short piece of ¼ in. dowel. This rests upon a single thickness of tin, D, which is cut into a strip ⅜ in. wide and about 1¼ in. long. In the bottom of C are two holes just large enough to allow the screws E and F to pass through. The wires, A and B, pass from the binding-posts, X and Y, through small holes burned through the sides of the box, and are fastened under the screw-heads. The whole box is screwed to the wooden base, which is 3 × 4 × ⅞ in., by the screws, E and F. D should have enough spring in it to raise itself and G when the pressure of the finger is removed. The circuit will be closed only when you press the button.

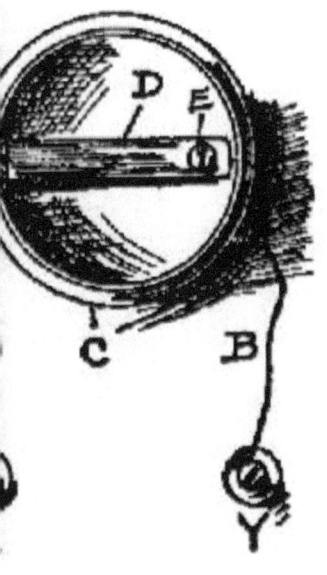

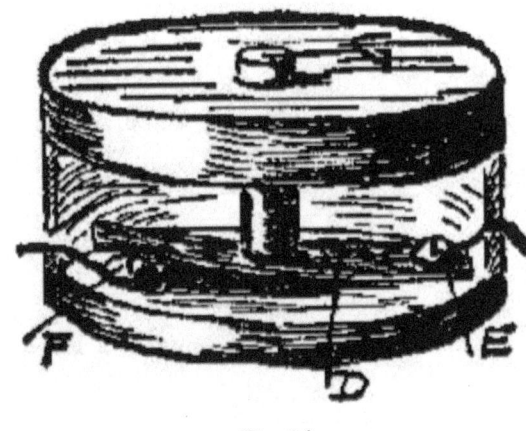

Fig. 16.

Fig. 15.

[Pg 25]

APPARATUS 31.

47. *Push-Button.* Figs. 17, 18, 19. Fig. 17 shows a top view or plan of the apparatus. Fig. 18 is a sectional view; that is, we suppose that the button has been cut into two parts along its length and through the center line. Fig. 19 is an enlarged detail drawing of the underside of the spool, C. The same part is marked by the same letter in all of the figures.

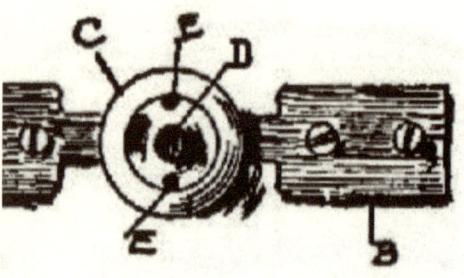

Fig. 17.

Fig. 18.

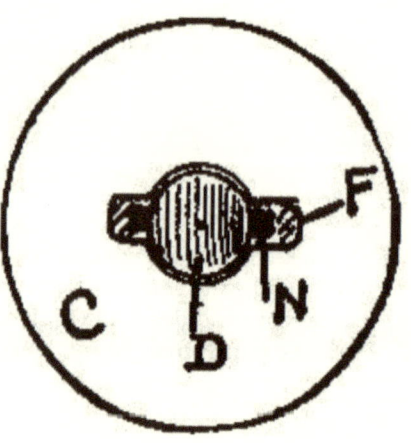

Fig. 19.

Saw an ordinary spool, C, into two parts. One-half of C will serve as the outside case for the button. The part to be pressed with the finger is a short length of ¼ in. dowel. To keep this from falling out of the hole in C, a short piece of wire nail, N, has been put through a small hole in its lower end. A slot, F, has been burned or cut into the underside of C, so that N can pass up and down in it when D is raised and lowered. The rod, D, rests upon A, one of the contacts. This is a straight piece of tin, cut as shown in Fig. 17, the narrow part being ¼ in. wide and 1¼ in. long. The wide part is ¾ in. wide and 1 in. long. The other contact, B, is the same size as A. A deep groove, a little over ¼ in. wide, is cut into the base so that the narrow part of B can be bent down below the end of A. The base shown is 4 × 2½ × ⅞ in. The spool, C, is fastened to the base by 2 screws or wire [Pg 26] nails put up through the base, their positions being shown by the dots at E, Fig. 17. X and Y, Fig. 18, are 2 screw binding-posts. It is evident that the current cannot pass from X to Y, unless the button, D, be pressed down so that the end of A will touch B.

APPARATUS 32.

48. *Sifter for Iron Filings.* Fig. 20. In making magnetic figures with iron filings, it is an advantage to have the particles of iron fairly small and uniform in size. A simple sifter may be made by pricking holes in the bottom of a pasteboard pill-box with a pin. The sifter may be put away with the filings in it, provided you turn it upside down.

Fig. 20.

APPARATUS 33.

49. *Sifter for Iron Filings.* Fig. 21. Punch small holes in the cover of a tin box with a small wire nail. If you have occasion to use sifters for other purposes, the different sizes can be made by using larger and smaller nails to punch the different tin covers. But one size of nail should be used for one sifter.

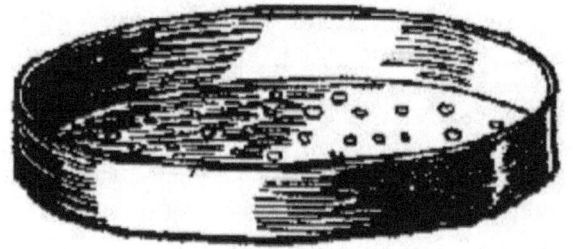

Fig. 21.

APPARATUS 34.

50. *Sifters* may be made by pricking holes in an envelope. A sifter with very small holes can be made of a piece of muslin cloth. This can be used in the form of a little bag, or a piece of it can be pasted over the open bottom of a pill-box.

[Pg 27]

APPARATUS 35.

51. To *Cut Wires, Nails, etc.* If you have no wire-cutters, or large shears, you can cut large or small wires by hammering them against the sharp edge of another hammer, an anvil, or a piece of iron. Do not let the hammer itself hit upon the edge of the anvil. The above process will make a V-shaped dent on one side of even large wires, or nails, when they may be broken by bending back and forth.

[Pg 28]

CHAPTER IV.

SWITCHES AND CUT-OUTS.

52. *Switches, Cut-Outs*. Where apparatus is to be used frequently, such as for telephone and telegraph lines, it pays to make your switches, etc., carefully. The use of these switches, etc., will be shown in the proper place. Their construction only will be given here.

APPARATUS 36.

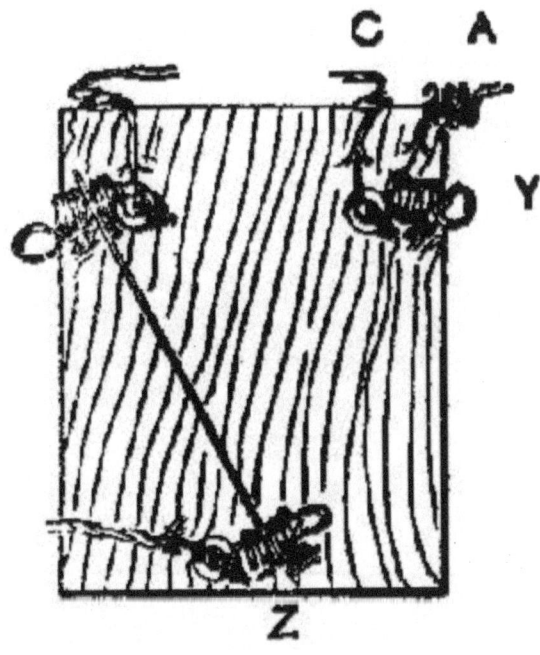

Fig. 22.

53. Cut-Out. Fig. 22. Details. X, Y, and Z represent 3 binding-posts like App. 42. These are fastened to a wooden base that is about 3 × 5 × ¾. The ends of the wires shown come from and go to the other pieces of apparatus. Q shows a stout wire or strip of 2 or 3 thicknesses of tin. Suppose we have an apparatus, as, for example, an electric bell, which we want to have ring when someone at a distance desires to call us. If we use a telephone or telegraph instrument we shall want to cut the bell out of the circuit as soon as we hear the call and are ready to talk. Suppose the current comes to us through the wire, A, Fig. 22. It can pass by the wire, C, through the bell and back to X. If we wanted simply to have the bell ring, the current could pass directly from X into the earth, or over a return wire back to the push-button at our friend's house. If, however, we are to use some other instrument, by lifting the end of Q out of X and pushing it into Y, the bell will be cut out, and the current can pass on wherever we need it.

[Pg 29]

APPARATUS 37.

54. *Cut-Out.* Fig. 23. The main features of this are like those of App. 36. The three binding-posts are like App. 46. Instead of a band of metal to change connections, as Q in App. 36, a stout copper wire is used. This can be easily changed from one of the upper binding-posts to the other, thereby throwing in or cutting out any piece of apparatus joined with the upper connectors.

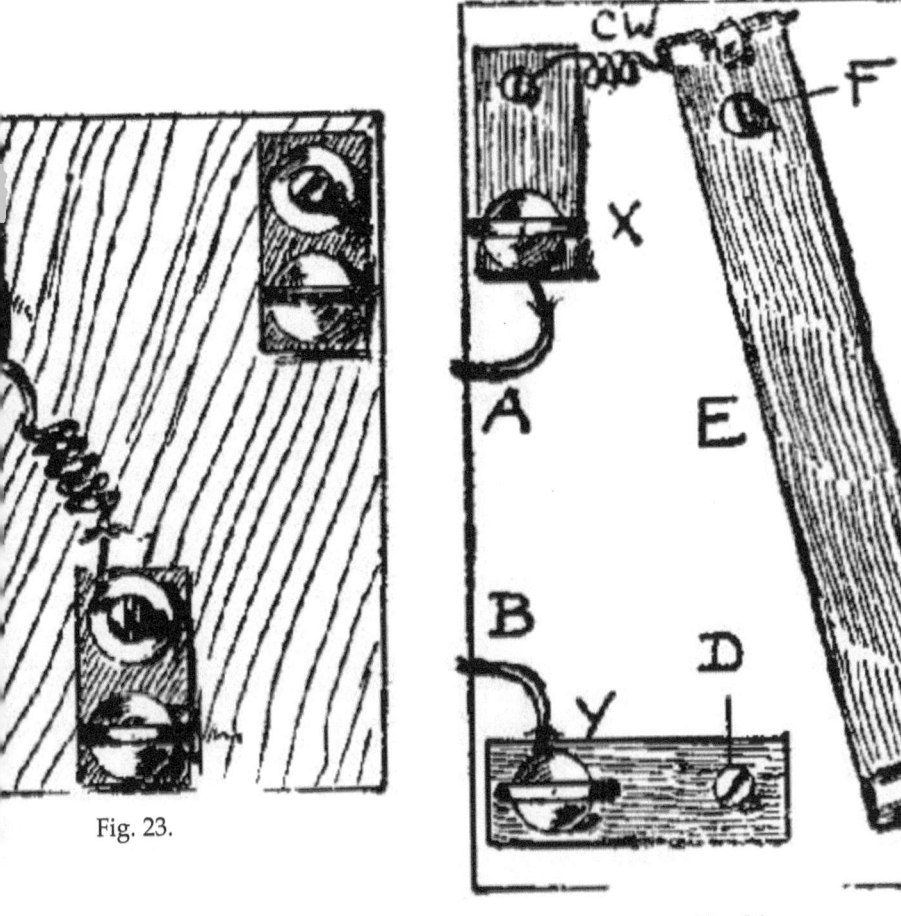

Fig. 23.

Fig. 24.

APPARATUS 38.

55. *Switch.* Fig. 24. This simple switch has but one contact point, D, which is a screw-head. This switch may be used anywhere in the circuit by simply cutting the wire carrying the current, and joining the ends of the wire to the binding-posts X and Y. The metal strip, E, is made of 2 or 3 thicknesses of tin. It is ⅝ in. wide and about 5 in. long, and presses down upon D, when swung to the left, thus closing the circuit. The short metal strips shown are ⅝ × 1¼ in. The upper strip is joined to the end of E by a coiled copper wire, CW. (See App. 50.) If the current enters by the wire, A, it will pass through CW, E, D and out at B. The strip E [Pg 30] is pivoted at F by a small screw. The base may be 3 or 4 × 5 × ⅞ in.

APPARATUS 39.

56. *Switch.* Fig. 25. By increasing the number of contact points and the wires leading from them, a switch may be made to throw in one or more pieces of apparatus. This variety of switch is useful in connection with resistance coils (Index). By joining the ends of the coils with the points 1, 2, 3, etc., more or less resistance can be easily thrown in by simply swinging the lever, E, around to the left or right. The uses of this will be again referred to.

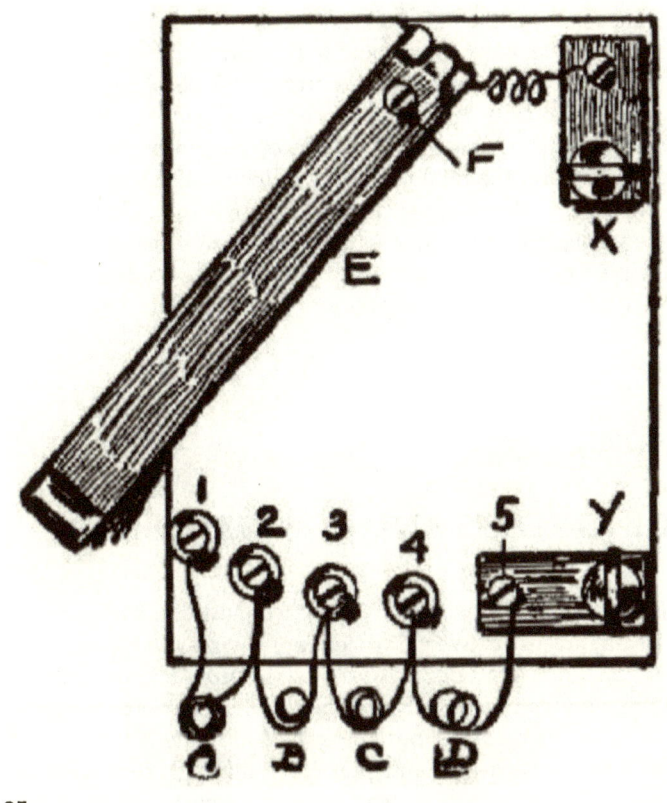

Fig. 25.

Details. The base of the one shown in Fig. 25 is 4 × 5 × ⅞ in. thick. The switch, E, is a band of 2 thicknesses of tin ⅝ in. wide. It is pivoted at F with a screw. To the end of E is fastened a copper wire, which leads to the upper binding-post, X (App. 46). The apparatus has 5 contact points, marked 1, 2, 3, etc. These consist of brass screws and copper washers. With F as a center draw the arc of a circle that has a radius of 4 in. Place [Pg 31] the screws 1, 2, etc., along this arc, and about ⅝ in. apart, center to center; that is, the screws are all 4 in. from F, and are, therefore, in the form of a curve.

The last screw forms a part of the binding-post, Y. Suppose 4 pieces of apparatus, marked A, B, C, and D, be connected with 1, 2, etc., as shown. These may be, for example, coils of wire to be used as resistance coils. If the current enters at X, it will pass along at E and be ready to leave at Y, as soon as E touches one of the contact points. If E be placed upon 1, the current will be obliged to pass through all of the coils, A, B, etc., before it can get to Y. In this case the resistance will be greatest. If E be now moved on to 2, only A will be cut out, and the total resistance reduced. By placing E upon 4, but one coil, D, will be in the circuit. When E is upon 5 the current will pass through the switch with practically no resistance. This is the principle upon which current regulators work. (Study resistance in text-book.) When E is in the position shown in Fig. 25 no current can pass.

[Pg 32]

CHAPTER V.

BINDING-POSTS AND CONNECTORS.

57. Binding-Posts are used to make connections between two pieces of apparatus, between two or more wires, between a wire and any apparatus, etc., etc. They are used simply for convenience, so that the wires can be quickly fastened or unfastened to the apparatus. There are many ways of making them at home. The following forms will be found useful and practical. Although some that are given are really connectors instead of binding-posts, we shall give them the general name of binding-posts.

APPARATUS 40.

58. Binding-Post. About the simplest form is a screw, or a nail with a flat head. The bare wire may be placed under the head of the screw or nail before forcing it entirely into the wood. This will keep the end of the wire in place, and another wire may be joined electrically to the first by merely touching it to the screw-head, or by placing it under the screw-head.

APPARATUS 41.

Fig. 26.

59. Binding-Post. Fig. 26. This consists of a screw and a copper washer or "bur." The screw is a "round-headed brass" one, ⅝ in. long, number 5 or 7. The copper burs are No. 8, and fit nicely around the screws. By using 2 burs instead of 1, several wires may be easily joined together at one point. Scrape the covering from the ends of the wires, and place them between the burs.

[Pg 33]

APPARATUS 42.

60. Binding-Post. Fig. 27. A coiled spring serves very well as a connector. One end should be fastened to the apparatus, as shown, by clamping it under a screw-head. The other end of the coil should be pulled out a little, away from the other turns, so that you can stretch the spring in order to put the bare ends of wires between the turns. Any number of wires placed between these turns will be pinched and electrically connected. The coil should be about ½ in. long and less than ½ in. in diameter. You can make a coil by tightly wrapping stiff iron wire around a pencil. The steel wire springs taken from old window-shades are excellent for this purpose. They may be cut into lengths with tinner's shears.

APPARATUS 43.

61. Binding-Post. Fig. 28. Two copper or tin strips fastened at one end by a screw, the upper strip being bent a little at one end, make a connector that is useful for some purposes, where you want to make and break the connection frequently. The bare end of the wire which belongs to the apparatus is fastened under the screw-head. The outside wire, or wires, to be connected are pushed between the strips of metal. Another way is to fasten the outside wire to a strip of metal about ½ in. wide, and then push this between the strips shown in the figure. The strips shown should be about ¾ in. wide and 1¼ in. long.

Fig. 27.

Fig. 28.

Fig. 29.

APPARATUS 44.

62. Binding-Post. Fig. 29. A combination made between App. 42 and 43 does well. Fasten a metal strip, [Pg 34] ¾ in. × 1¼ in., to the apparatus by means of a screw. The apparatus wire should be fastened under the screw-head. A short length of spring may be pushed upon the upright part of the strip, as shown. Into this you can quickly fasten the outside wires.

APPARATUS 45.

63. Binding-Post. Fig. 30. This makes a very simple and practical binding-post for home-made apparatus. It consists of a screw-eye, preferably of brass. The circle or eye should be about ⅜ or ½ in. in diameter. The thread on such a screw-eye will be about ½ in. long. Two copper burs are used to pinch the wires.

APPARATUS 46.

64. Binding-Post. Fig. 31. This consists of a screw, screw-eye, bur and a metal strip, ¾ × 1¼ in. The apparatus wire should be fastened under the screw-head. Any outside wires which are to be joined to the apparatus should be clamped under the bur by turning the screw-eye. A small hole should be made in the wood before putting in the screw-eye. (See App. 25.) Do not turn the screw-eye too hard, or you will spoil the thread made in the wood.

Fig. 31.   Fig. 32.

Fig. 30.

APPARATUS 47.

65. Binding-Post. Fig. 32. The size of the bolt used in this form of binding-post will depend somewhat upon the thickness of the base of the apparatus. In general, a ¾ or ⅞ in. base should be used where screws or [Pg 35] screw-eyes are necessary. With this kind (Fig. 32) a thin base can be used. The head is shown counter-sunk into the bottom of the base. This is not necessary, provided at least 3 heads are placed far enough apart to form legs for the apparatus to stand on. Strips of wood may be nailed upon the underside of the base to make room for the heads in case they are not used as legs. The wires should be pinched between the nut and the copper bur shown. If the bolt is too large for a bur, an iron washer may be used. A washer may be made of tin, or two nuts may be used.

APPARATUS 48.

Fig. 33.

66. Binding-Post. Fig. 33. This is a suggestion for a combination of App. 44 and 47. It is useful in school apparatus. Wires may be permanently fastened on the right, under the nut, and a spring, as in App. 44, may be slipped on the metal strip at the left, which is held under the head of the bolt.

APPARATUS 49.

67. Mercury Connector. A cup of mercury may be used as a connector. Make a small hole about ¼ in. in diameter and depth, in a piece of wood, and place 2 or 3 drops of mercury in this. The ends of wires dipped in this will be electrically connected.

APPARATUS 50.

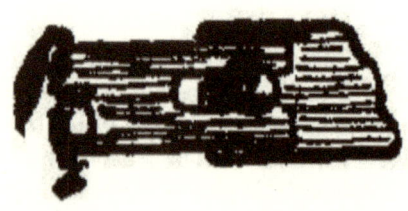

Fig. 34.

68. Connector. Fig. 34. This shows how a wire may be fastened to one end of a short strip of tin. At the other end of the strip a slot is cut. This may straddle the body of a screw, or when left plain may be used to slip between the two metal strips shown in App. 43.

[Pg 36]

APPARATUS 51.

69. Binding-Post. Fig. 35. The ends of two or more wires may be quickly joined electrically by placing them between the nuts of a short bolt. By using 3 nuts the bolt will more easily connect a large number of wires.

Fig. 35.

Make Additional Notes and Sketches Here.

[Pg 37]

CHAPTER VI.

PERMANENT MAGNETS.

70. *Permanent Magnets* may be made in many ways and from many different kinds of steel. The steel used for needles, watch and clock springs, files, cutting tools, etc., is generally of good quality, and it is already hard enough to retain magnetism. (See Retentivity in text-book.)

APPARATUS 52.

71. *Bar Magnet.* A straight magnet is called a bar magnet. Magnetize a sewing-needle. For some experiments a needle-magnet, as we may call it, is better than a large magnet.

APPARATUS 53.

72. *Bar Magnet.* A harness-needle, which is thicker and stronger than a sewing-needle, makes an excellent bar magnet.

APPARATUS 54.

73. *Bar Magnet.* For long slim magnets use a knitting-needle. Some knitting-pins, as they are sometimes called, break off short when bent, but most of them will bend considerably before breaking. These slim magnets are excellent for the study of Consequent Poles. (See text-book.)

APPARATUS 55.

74. Flexible Bar Magnets. It is often necessary to have flexible magnets so that they may be bent into different shapes. These may be made from watch or clock springs, as such steel, called spring steel, will straighten out again as soon as the pressure is removed from it. Corset [Pg 38] steels, dress steels, hack-saw blades, etc., make good thin flexible bar magnets.

APPARATUS 56.

75. Strong Bar Magnets may be made from flat files. The handle end may be broken off so that the two ends of the file shall be nearly alike in size. These should be magnetized upon an electro-magnet.

APPARATUS 57.

76. Compound Bar Magnets are made by first magnetizing several thin pieces of steel, and then riveting them together so that their like poles shall be together, and pull together. To make a small compound bar magnet, magnetize several harness-needles, or even sewing-needles, and then bind them into a little bundle with all the N poles at the same end. Melted paraffine dropped in between them will hold them together. Rubber bands may be used also, or, if but one end is to be experimented with, the points may be stuck into a cork, and the heads used to do the lifting.

APPARATUS 58.

77. Small Horseshoe Magnets may be made from needles or from other pieces of steel used for bar magnets. They should be annealed (App. 21) at their centers at least, so that you can bend them into the desired shape. In the case of bright needles, like harness-needles, the part annealed will become blackened. If you heat the center only, and the ends remain bright for about ½ inch, you will not need to harden the needle again. It is an advantage to have the center of the magnet a little soft, as it is not then liable to break. The ends alone may be hardened by holding the bent portion away from the candle or gas flame, while heating the ends. The bent steel should be magnetized by drawing its ends across the poles of a horseshoe magnet.

[Pg 39]

APPARATUS 59.

78. Flexible Horseshoe Magnets may be made of thin spring steel. The distance between the poles can be regulated at will by bending the steel more or less. The poles may be held at any desired distance apart by thread or wire, which should be wound around the legs of the magnet a little above the poles. This will keep the steel from straightening out.

APPARATUS 60.

79. Horseshoe Magnet. Fig. 36 and 37. Magnetize two harness-needles, and stick them into a cork so that the poles shall be arranged as shown. The distance between the poles can be regulated to suit. This forms a very simple and efficient magnet, with the advantages of a real horseshoe magnet.

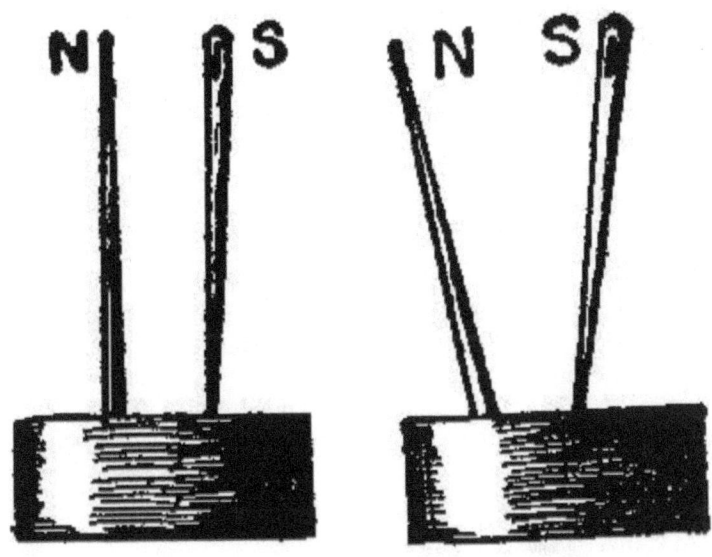

Fig. 36.   Fig. 37.

APPARATUS 61.

80. Armatures. All home-made magnets should be provided with armatures, or keepers. These are made of soft iron on the regular magnets, and tend to keep the magnet strong. (See text-book.) For the bar magnets described, a piece of sheet-tin, upon which to lay them, is all that is needed for an armature. The lines of force will pass through this. For the horseshoe magnets described, strips of tin, soft iron wires, or even a wire nail placed across the poles will greatly aid in keeping in the strength. The little magnets should not be dropped or jarred. (Study the theory of magnetism in text-book.)

[Pg 40]

CHAPTER VII.

MAGNETIC NEEDLES AND COMPASSES.

81. *Magnetic Needles and Compasses* consist chiefly of a short bar-magnet. When used to tell the directions, north, east, etc., the apparatus is generally called a compass. When we speak of the "needle," we really mean the compass-needle. The little magnet may be almost any piece of magnetized steel, provided it is arranged so that it can easily swing around. There are several ways of supporting the compass-needle. It may rest upon a pivot, it may be hung from a fine thread, or it may be floated upon water with the aid of a cork, etc.

82. *Uses*. We all know that compasses are used to point to the north and south, and we speak of the "points of the compass." This, of course, is the most important use of the compass, and it has been known for centuries. In the laboratory it is used to *show or detect the presence of currents of electricity*, and, in connection with coils of wire, it may show the relative strengths of two currents, etc. When used for such purposes it generally has special forms and sizes. (See Galvanometers and Detectors.)

APPARATUS 62.

83. *Compass*. An oily sewing-needle will float upon the surface of water, when it is carefully let down to the water. A little butter may be rubbed upon the previously-magnetized needle to make it float better.

Fig. 38.

[Pg 41]

APPARATUS 63.

84. Compass. Fig. 38 shows a magnetized sewing-needle floated upon a cork. The needle may be permanently fastened to the cork with a few drops of melted paraffine.

APPARATUS 64.

85. Compass. Fig. 39. With a sharp knife make a cut part way through a flat cork. Into the cut push a short length of magnetized watch-spring. In the illustration the spring is shown partly removed from the cut. Float the cork.

APPARATUS 65.

86. Compass. Fig. 40. Stick a pin, P, into a pasteboard, cork, or wooden base, B. Bend a piece of stiff paper double, as shown, and then stick through it, on each side, a magnetized sewing-needle, S N. The north poles of the needles should be at the same end of the paper. Why? Balance the paper upon the pin-pivot, and see it fly around to the north and south.

Fig. 39.

Fig. 41.

Fig. 40.

APPARATUS 66.

87. *Compass.* Fig. 41. It is an advantage to have a magnetic needle that is always ready for use. *The support* is made by driving a pin through the *top* of a wooden pill-box, which should be about 1¾ in. in diameter. This gives plenty of room under and around the needle. If the pin be left too long, it will not be possible to put the bottom and top of the box together when you want to put [Pg 42] the compass away. Cut the pin off (App. 35) at the right length, so that the magnetic needle can be safely put away in the closed pill-box.

88. *The "Needle,"* that is the short bar magnet, may be made of watch-spring. As the spring is already quite hard and brittle, it may be easily broken into desired lengths. It is always better to make 3 or 4 needles at a time, as some will swing more easily than others, and time will be saved in making them. Break off 3 or 4 pieces of thin spring, each about 1½ in. long. Bend them as in Fig. 42. A good dent, not a hole, should be made at the center of each to keep them upon the support or pin-point. A "center punch," not too sharp, is the best tool to use, but a slight dent may be made with a sharp wire nail, provided the watch-spring is first annealed or softened. (See App. 21.) Do not place the spring directly upon iron or steel when making the dent, as these might injure the point of the punch, and the dent would not be deep enough. Fig. 42 shows a good way to make dents in steel springs. Place 2 or 3 layers of copper or lead between the anvil and the spring. A hammer or hatchet will do for the anvil. As the copper will give easily, a good dent may be made by striking the punch or nail with a hammer. If the spring has been annealed before denting it, it should be hardened again (App. 21) before magnetizing it, so that it will retain magnetism well. (See Residual Magnetism in text-book.)

Fig. 42.

89. Balancing. After a dent has been made, place the spring upon its support so that the pin-point shall be in the dent. It will, no doubt, need balancing. If one end is [Pg 43] but *slightly* heavier than the other, the spring may be balanced by magnetizing it so that the lighter end shall become a north pole. This will then tend to "dip" and make the needle swing horizontally. If one end is *much* heavier than the other, it should first be magnetized and then balanced by cutting little pieces from the heavier end with tinners' shears, or by weighting the lighter end with thread, which may be wound around it. The finished compass-needle should swing very freely, and should finally come to rest in an *N* and *S* line after vibrating back and forth several times.

APPARATUS 67.

90. Glass-Covered Compass. A perspective view of this apparatus is shown in the tangent galvanometer. (See Index.) The outside band, *E*, is made of thick paper, 1 in. wide, and with such a diameter that it just fits around the glass. In this model, the glass from an old alarm-clock was used, it being 4 in. in diameter. Four pasteboard strips were sewed to the inside of the paper band *E*. They were made ⅞ in. long, so that the glass, when resting upon them, would be near the top of *E*.

The needle should be not over 1 in. long, if it is to be used in the galvanometer. A long slender paper pointer should be stuck to the top of the needle. Be careful to have the combined needle and pointer well balanced, so that it will swing freely. A circle graduated into 5-degree spaces should be fastened under the needle.

91. Astatic Needles. In the magnetic needles so far described, the pointing-power has been quite strong. By pointing-power we mean the tendency to swing around to the N and S. In App. 65 the 2 needle magnets had considerable pointing-power, because they helped each other. For some experiments in electricity a magnetic needle is [Pg 44] required which has but *little* pointing-power; in fact, to detect the presence of very feeble currents by means of the needle, the less the pointing-power the better. Can you think of any way to arrange App. 65 so that it shall have very little pointing-power?

APPARATUS 68.

92. Astatic Needle. Fig. 43. Turn one of the needle magnets of App. 65 end for end, so that the N pole of one shall be at the same end of the paper as the S pole of the other. You can see that by this arrangement one needle pulls against the other. The magnetic field still remains about the little magnets, otherwise this combination would be of no value in the construction of galvanometers. The more nearly equal the magnets are in strength, the less the pointing-power of the combination.

Fig. 43.

Fig. 44.

APPARATUS 69.

93. Astatic Needle. Fig. 44. Magnetize two sewing-needles as equally as possible, by rubbing them over the pole of a magnet an equal number of times. Remove the covering from a piece of fine copper wire, say No. 30, and use the bare wire to wind about the needles, as shown. Be sure to place the poles of the little magnets as in the Fig. This combination may be supported by a fine thread. It is used for Astatic Detectors. (See Index.)

[Pg 45]

CHAPTER VIII.

YOKES AND ARMATURES.

94. Yokes are used to fasten two straight electro-magnets together to form a horseshoe electro-magnet. The reasons for using them should be understood. Soft iron should be used for yokes and armatures, as this is the best conductor of lines of magnetic force. Sheet-tin is made of thin iron, which is coated with tin. (Try a magnet upon a tin can.) This soft iron is very easily handled, bent, and punched, and is very useful for many purposes. The tin from old tomato cans, cracker boxes, etc., is just as good as any. The method of making your yokes will depend entirely upon the tools at your command. Several ways are given. Y, Fig. 47, shows the position of the yoke.

APPARATUS 70.

95. Yoke. For the experimental magnets (App. 89) a fairly large yoke is required in order to have the magnets far enough apart. If you have only a nail punch (App. 26) with which to make holes in tin, you will be obliged to punch but one thickness at a time. (See method of punching sheet-metal, App. 26.) Cut 5 or 6 pieces of the tin, 3¼ × 1 in. With a center punch (tools) or sharp-pointed nail make small dents (2 in. apart) in each piece to mark the places where the holes are to be punched. Punch 5/16 in. holes in each piece. If you do this carefully, the holes in the different pieces will match, and the bolts can be pushed or screwed into these. When screwing in the bolt magnets turn them by their heads; do not pinch the coils, as this loosens the wire.

[Pg 46]

If you have a good punch, it is better to make the yoke as in App. 27, instead of using separate pieces of tin.

APPARATUS 71.

96. Yoke. Fig. 45 and 46. Cut a strip of tin 6 in. long by 3¼ in. wide. Bend one end of it so that it will lap over ¾ in. (Fig. 46); hammer it down gently, then bend this over and over until the whole tin is used. The final result will be a flat roll, 3¼ by about 1 in. This should be hammered flat.

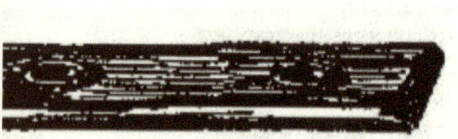

Fig. 45.

Fig. 46.

If you have the tools it is easy to drill two 5/16 holes in this strip. They should be 2 in. apart; that is, 2 in. from the center of one to that of the other. Start the holes with a center punch.

97. If you have no way of drilling the holes, they must be punched. (See App. 27.) This will make the strip bulge out on the underside around the holes. This bur, or most of it, should be filed off. (See App. 79 for method of filing thin pieces of metal.) The resulting yoke may be held firmly to the magnets by the use of 2 extra nuts, as in Fig. 67. Remember that the magnets must be held *firmly* in the yoke.

APPARATUS 72.

98. Yoke. The best way of making this, of course, is to cut a piece of bar-iron the right size. For 5/16 bolts the strip of wrought iron should be about ¾ in. wide and 3/16 or ¼ in. thick. Any blacksmith can make this and punch or drill the holes. If taps and dies (tools) are at [Pg 47] hand, the hole may be drilled and tapped to fit the thread on the bolt. It is very easy to make good looking apparatus if you have, and can use, a whole machine shop full of tools. The lengths of yokes will depend upon the special uses to be made of them.

APPARATUS 73.

99. Yoke. Fig. 47. The yoke, Y, is a part of a carriage. This can be bought at a blacksmith's. The holes are already in, but it may require some filing before the nuts of the bolt magnets will fit down firmly.

Fig. 47.

APPARATUS 74.

100. Tin Armatures may be made by bending together 5 or 6 thicknesses of tin. Different forms of tin armatures are shown under telegraph sounders; these should have a hole punched at the center; through this is put a screw. The length of the armature will depend upon the distance the magnets are placed apart; they should be about ¾ in. wide.

APPARATUS 75.

101. Nail Armatures. Fig. 48. A nail, N, placed through a piece of wood, A, will serve as a very simple armature. To make it a little heavier, if necessary, a piece of annealed iron wire, W, may be wound around N. Care should be taken to have the two parts fairly alike in size and weight.

Fig. 48.

APPARATUS 76.

102. *Wire Armatures.* Fig. 49. Annealed iron wires make good armatures. The short lengths of wire should be straightened (See App. 28) before binding them [Pg 48] into a bundle. They may be held together with thread or paraffine, until they are in place, as, for example, in a wooden piece, A, Fig. 49. The bundle of wires should fit snugly into the hole made through A, and the wires should be bound together at each end with wire.

Fig. 49.

APPARATUS 77.

103. *Trembling Armature.* Fig. 50. Armatures to be used upon electric bells, automatic current interrupters, buzzers, etc., may be called trembling armatures. They may be made entirely of sheet-tin. The part, F, which gives it the spring, should be about ⅝ in. wide. Its length will depend upon the particular apparatus to be made. It is made of 2 thicknesses of thin tin. See Fig. 50 for dimensions. The part N projects beyond L. This may be used to tap against a regulating screw, or to fasten a hammer on for an electric bell. The part, L, should have about 4 layers of tin on each side of F, and it should pinch F tightly.

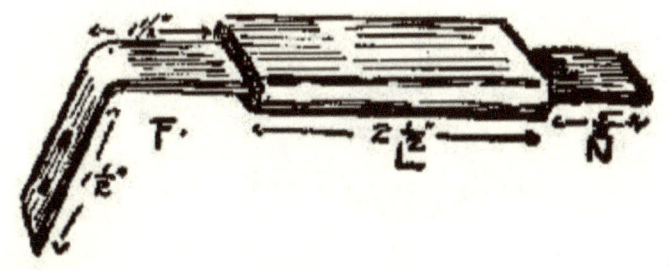

Fig. 50.

APPARATUS 78.

104. Trembling Armature. Fig. 51. When very rapid motions are desired in a trembling armature, App. 77 will be a little heavy. A light and quick-acting armature can be made of sheet-tin. The exact dimensions will depend upon the use to be made of it, but you will find the following a guide. Cut the part, B, E, out of [Pg 49] *thin* tin. The covers and bottoms of tin cans are thinner than their bodies. The narrow part, B, should be about ¼ in. wide and 2 in. long for a small apparatus, while E may be ¾ in. square. Through E is a screw, which holds it firmly to a wooden piece, D, about ¾ in. square. The part, E, can be made longer than its width, so that two screws can be used; this will keep A from jarring up or down.

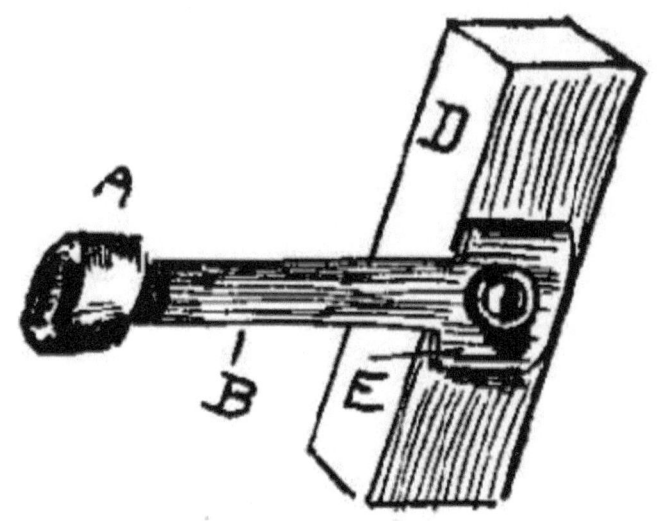

Fig. 51.

APPARATUS 79.

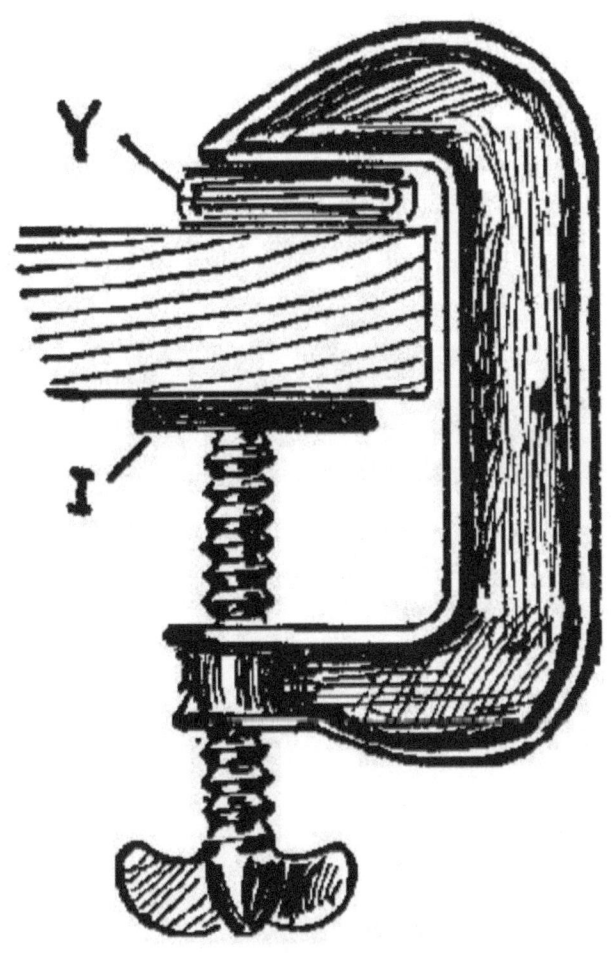

Fig. 52.

105. To File Thin Metal Strips. Fig. 52. When sheet-metal is punched by the methods usually employed by boys, a bulge or bur is made on the underside around the hole. If this bur be hammered to flatten it, the hole is distorted and made smaller. It is better to file the bur [Pg 50] down, at least part way. It is not convenient to file a piece of thin metal when it is held in a vise. It is better to use either a metal or a wooden clamp, as shown in Fig. 52; then the filing can be quickly and easily done. Y is the yoke to be filed. It is well to place a piece of metal, I, between the table and the end of the screw.

APPARATUS 80.

106. Clamp. Fig. 53. If you have no clamp to hold metal strips while filing them, you can put a screw, S, through one hole to hold the strip down fairly tight. Drive a nail, N, behind the strip. This will keep it from turning while you file the free end.

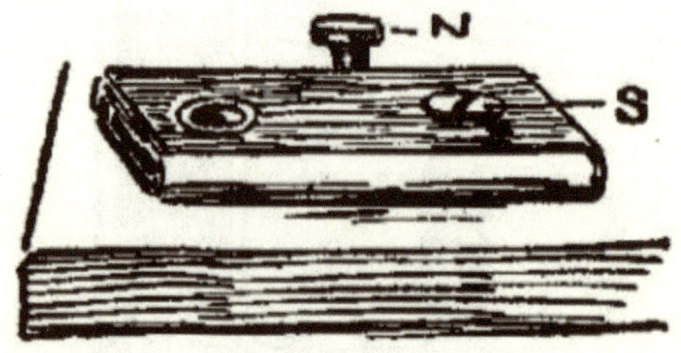

Fig. 53.

Electro-Magnetic Armatures. A description of this form of armature is given in the chapter on electric motors. (See Index.)

[Pg 51]

CHAPTER IX.

ELECTRO-MAGNETS.

107. Electro-Magnets are absolutely necessary in the construction of most pieces of electrical apparatus. There are several ways of making them at home. To quickly make a good-looking one, a winder (App. 93) is required. We shall divide our electro-magnets into four parts: *Core, washers, insulation, and coil.*

Of course, you know that when a current of electricity passes through a wire, a magnetic field is produced around the wire. A coil of wire, or helix, has a stronger field than a straight wire carrying the same current, because each turn or convolution adds its field to that of the other turns. By having the center of the helix made of iron, instead of air, wood, or other non-magnetic bodies, the strength of the magnet is greatly increased. This central core may be fixed permanently in the coil, or be removable. For our purposes fixed cores are just as good as movable ones, and the coils are easily wound upon them.

When wire is wound by hand from a spool into a coil, or around a core, it soon becomes twisted and tangled. Make a winder. This will keep the wire straight and save much time.

APPARATUS 81.

Fig. 54.

108. Electro-Magnet. Fig. 54. Drive a nail into a board so that it will project about ¾ of an inch. A soft, or wrought-iron, nail is best, but a short, thick wire-nail will do. If you do not have a thick nail, use an iron screw. Wind 3 or 4 layers of insulated copper wire around it, and fasten the bare ends of the wire down with bent pins. Number 24 wire will be found a good size for [Pg 52] experimental purposes. Touch the wires leading from the battery to the ends of the coil, and see if the nail will lift pieces of iron.

109. Note. Always leave at least 6 in. of wire at the ends of all coils and windings. This is needed for connections and repairs, as the wire is liable to get broken at any time around the binding-posts.

110. Note. After you have wound wire upon a core or spool, keep it from untwisting by taking a loop or hitch around it with the wire. Fig. 55 shows how this is done. Pull the end of the wire enough to make the loop stay in place.

APPARATUS 82.

111. Electro-Magnet. Fig. 56. Cut annealed iron wire into pieces, 3 inches long, straighten them (App. 28), and tie them with thread into a bundle about 5/16 in. in diameter. Melted paraffine run in between the wires will hold them in together, but stout thread will do. Wind 3 or 5 layers of No. 24 insulated copper wire upon the soft iron core. This is useful for simple experiments, and this idea may be applied to magnets to be used in pieces of apparatus. Hold the bundle of wires in a vise, and file the ends smooth, before winding on the wire. Paraffine should be used to hold the turns of insulated wire together.

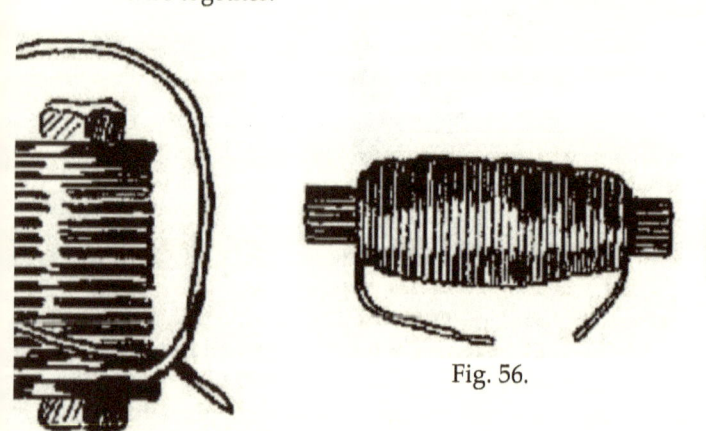
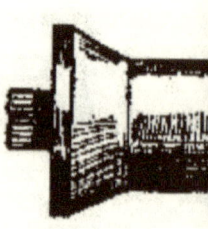

Fig. 55.

Fig. 56.

Fig. 5

APPARATUS 83.

112. *Electro-Magnet.* Fig. 57. An electro-magnet with a *removable core* may be made by winding the wire [Pg 53] on a spool. The core is made, as in App. 82, of soft iron wires, bound together with stout thread. A bolt may be used instead of the wire, but the wire loses its magnetism much quicker than a soft steel bolt would. (Study residual magnetism.) This magnet is strong enough for many purposes, but the wire is too far from the core, on account of the thickness of the wood, to make it efficient. The wire may be wound on by hand, but a winder (App. 93) will do much better and quicker work.

APPARATUS 84.

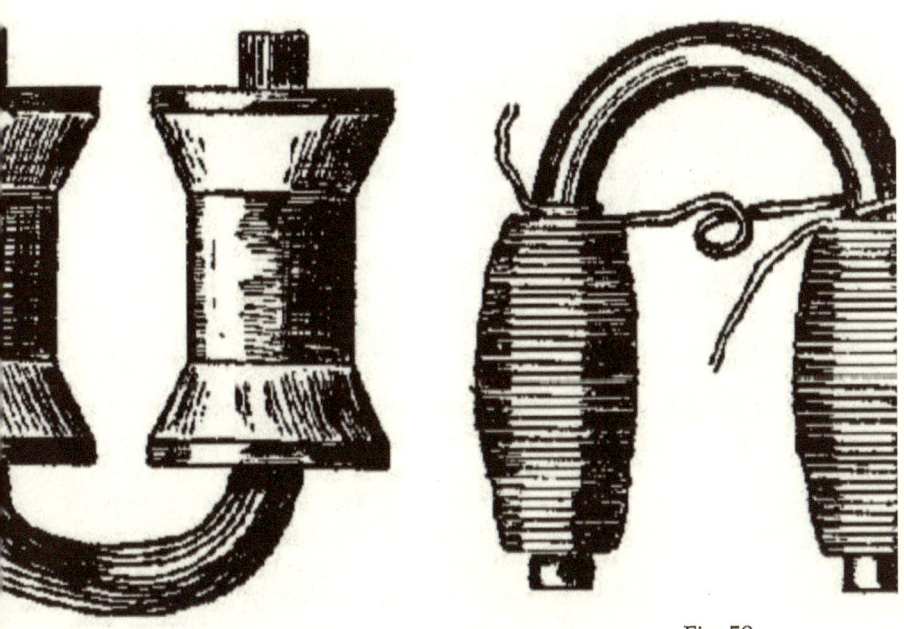

Fig. 58.   Fig. 59.

113. *Horseshoe Electro-Magnet.* Fig. 58. Bend soft iron wires, and make a bundle of them. If you wish to wind the wire around spools, the bundle cannot be very large. It will be found best to make the bundle about ⅜ in. in diameter, and not to use the spools. Strong paper should be wrapped once or twice around the legs of the horseshoe, and the insulated wire, say 4 layers, can then be wound directly upon this. (See § 115 for method of making connection between the coils.) It is a little troublesome to wind wire upon a horseshoe like this, and for App. 85. Spools are handier, because each can be wound separately, and then be slipped in place. The ends of the horseshoe should be filed smooth.

[Pg 54]

APPARATUS 85.

114. *Electro-Magnet.* Fig. 59. An ordinary iron staple is useful as the core of a small magnet. One like this is shown also in Fig. 94, used as a telegraph sounder. It takes some time to wind 4 layers of wire on to each leg of the staple, so be sure to see § 115 about the method of winding. In Fig. 59 the half-hitches (§ 110) are not shown. Coat the finished coils with paraffine.

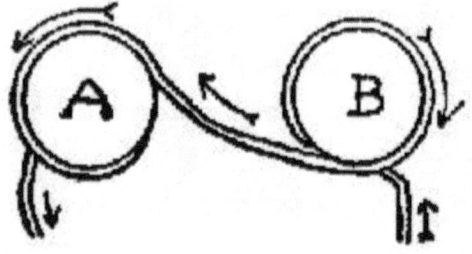

Fig. 60.

115. *Method of Joining Coils.* Fig. 60. If A and B represent the two cores of a horseshoe electro-magnet, the coils must be joined in such a manner that the current will pass around them in opposite directions, in order to make them unlike poles. The current is supposed to pass around B, Fig. 60, in the direction taken by clock hands, while it passes around A in an anti-clockwise direction. The inside ends, § 123, of the coils may be twisted together, or fastened under a screw-head. In Fig. 60 one coil is shown to be a continuation of the other.

APPARATUS 86.

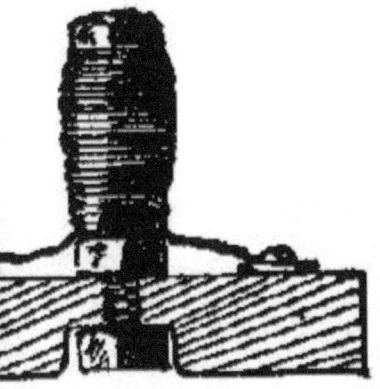

Fig. 61.

Fig. 62.

116. *Electro-Magnet.* Fig. 61. Wind 6 layers of No. 24 or 25 insulated copper wire around a 5/16 machine-bolt that is 2½ in. long. Fig. 61 shows one method of holding the bolt solidly in an upright position, so that magnetic figures can be easily made and the magnet studied. Two nuts are used, the lower one being counter-sunk, so [Pg 55] that the base will stand flat upon the table. This bolt is shown without washers (§ 119), and will do fairly well to show the action of electro-magnets. The ends of the wire should always be left 5 or 6 in. long, and be led out to binding-posts. The coil may be held in place, and its turns kept from untwisting by coating it with paraffine. The base may be of any desired size.

APPARATUS 87.

117. *Electro-Magnet Core.* Fig. 62. This shows another method of fastening a bolt-core in an upright position. This is done without the use of two nuts. A strip of tin, *T*, 1 in. wide, is punched and slipped onto the 5/16 bolt before the nut is screwed on and the coil wound. This is fastened to the base by screws, *S*. Washers, *W*, are here shown. (See § 119 for washers.)

APPARATUS 88.

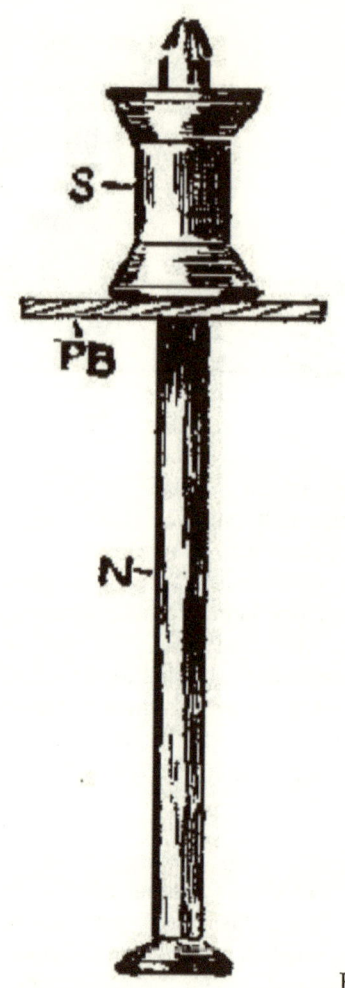

Fig. 63.

118. Bolt Electro-Magnets are easy to make, according to the directions given, and they are, when finished, more like the regular purchased magnets than any of the other forms described. With proper batteries (App. 3, 4, etc.,) they can be used for a great variety of purposes, as will be seen. There are many forms of bolts in the market, but the ordinary "machine bolt," 5/16 in. in diameter, is best for our purposes. The ones 2 and 2½ in. long are used.

119. Washers or coil ends are used on the bolt magnets so that considerable wire can be wound on closely and evenly. These are made out of thick pasteboard, which cuts smoother if it has been soaked in melted paraffine. Unless you know how, you will find it a hard job to make the hole in the exact center of the washer. The method of easily making washers is illustrated in Fig. 63.

First place a spool (the end of which is ⅞ or 1 in. in [Pg 56] diameter) upon the table, and lay the pasteboard upon this. Push a large round nail through the pasteboard into the hole in the spool. The nail should be nearly as large as the hole. Use the large nail as a handle, and with the shears cut around the edge of the spool end. Cut the washer as round as possible, and be careful not to cut into the spool.

The holes in the washers will be a little smaller than the 5/16 bolt. This will make the washers hold tightly to the bolt when you force them on. Fig. 64 shows the bolt-core, with the washers in place. If you cannot get a large nail, a lead-pencil, or sharpened dowel, will do to force through the pasteboard.

120. Insulation of Cores. While the covering on the wire would probably be all that is necessary to thoroughly insulate the coil from the core, it is better to wind a layer or two of paraffine paper around the bolt (Fig. 65) before winding.

121. The Coils of wire to be used upon the bolt-cores should be put on with the winder (App. 93). For all ordinary purposes No. 24 or 25 single or double cotton covered copper wire will do. It is better to put on an even number of layers. The winding (See Fig. 70) begins at the nut-end of the bolt, and by using 6 or 8 layers of wire, [Pg 57] instead of 5 or 7, both coil ends will be at the same end of the bolt.

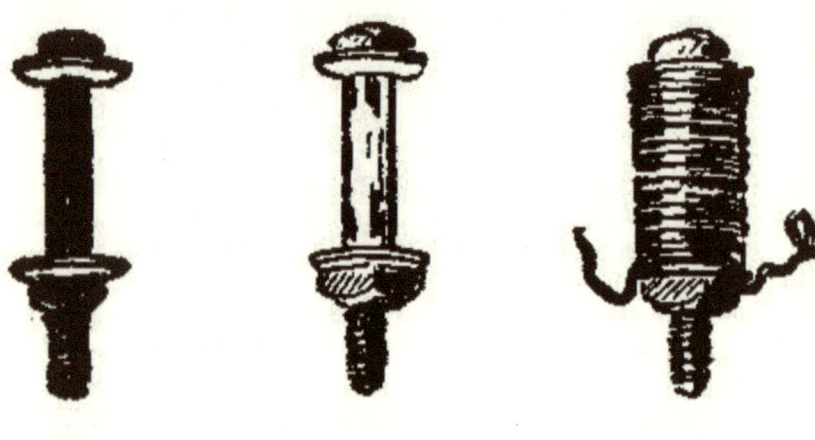

Fig. 64. Fig. 65. Fig. 65½.

122. Method of Winding the Coils. The winders used for bolt magnets are described in App. 91, etc. We shall suppose that the washer, W, Fig. 70, and the insulation, I, are upon the bolt before screwing it into the winder-nut, $W N$. Make a pinhole, $P H$, in the right-hand washer, as near the bolt-nut, $B N$, as possible. Stick about 6 in. of the wire through $P H$, and wind this end around $W N$, as shown, to hold the wire. The supply of wire should be upon a spool slipped onto some stationary rod (App. 23), so that you can give your entire attention to winding. Begin to turn the winder slowly at first. Turn the handle towards you when it is at the bottom, as in Fig. 70; that is, if you look at it from the side, turn the handle clockwise. Let the wire slip through your left hand as the turns are made, and guide it so that the turns will be close together. If they go on crooked, unwind at once, then rewind properly. You can guide the wire best by holding your left hand about 8 or 10 inches from the bolt. As soon as you reach the left side or head end of the bolt, feed the wire towards the right. If at any time the layers become rough on account of one turn slipping down between turns of the previous layer, fasten a piece of paraffine paper around the coil as soon as the imperfect layer is completed. Wind on 8 layers, and count the number of turns in one or two of them, so that you can tell about how many turns in all you have around the core. Make a "half-hitch" (see § 110) with the wire when the last layer is finished, to keep it from unwinding, and leave a 6 in. end.

The coil should be protected by fastening around it a piece of dark-colored stiff paper. Paraffine paper is good for this purpose. With a little practice you will be able [Pg 58] to rapidly and neatly wind on the wire. The winder-nut, *W N*, must hold the bolt solidly to keep it from wobbling.

123. We shall call the starting end of the wire which passes through *P H*, the *inside end*, and the end of the last layer the *outside end*. This can pass out between the washer and the paper covering.

APPARATUS 89.

124. *Experimental Horseshoe Electro-Magnet.* Fig. 66. Among the most useful pieces of apparatus for home use, is a good horseshoe electro-magnet. Fig. 66 shows a very convenient and practical form. With this, alone, can be shown all the principles of telegraph sounders, electric bells, etc. They are excellent for making magnetic figures (See text-book). You are supposed to be looking down on the App. in Fig. 66. The bolts are 2 in. apart center to center.

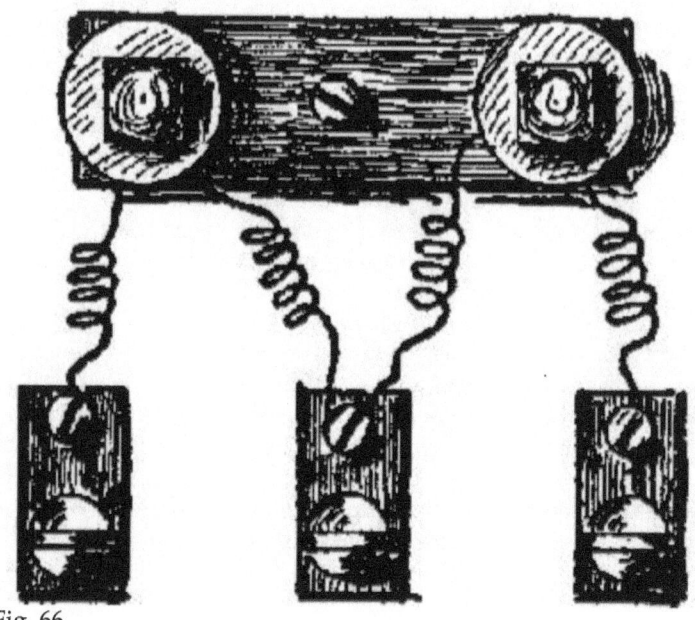

Fig. 66.

The bolt magnets are fully described in App. 88; *the binding-posts*, as App. 46; *the yoke*, as App. 71; the method of *fastening to the base*, as App. 90; the *base* is 5 × 4 × ⅞ in.; the magnets are made of 5/16 bolts, 2½ in. long.

125. To Join the Coils, fasten the two *inside* ends (§123) of the wire to a middle binding-post, and carry the *outside* ends to the two outside binding-posts. In this way [Pg 59] you can use either magnet alone, if desired (See experiments in text-book), or change the polarity at will by changing the connections. (See § 115 and 123.)

APPARATUS 90.

126. Fastenings for Electro-Magnet. Fig. 67. When both electromagnets are to be permanently fastened to a base, especially if tin yokes are to be used, as in App. 89, it is best to use a nut on each side of the yoke. It is important to have a perfectly tight connection between bolt and yoke. Several ways of fastening the bolts and yokes are shown; but it will be found best to cut holes in the base for the lower nuts, and to screw the yoke directly to the base. This makes a solid and pleasing arrangement. For the experimental magnets (App. 89) make the yoke 3¼ in. long, and place the magnets 2 in. apart center to center.

Fig. 67.

[Pg 60]

CHAPTER X.

WIRE WINDING APPARATUS.

APPARATUS 91.

127. Winder. Fig. 68. In case you do not have any means of making a smooth hole for the "bearings" of the winders of App. 93 and 94, you can use a spool for the purpose. *B* is the end of a piece of board about 1 in. thick, 3 in. wide, and 6 in. long. The spool, *A*, is laid upon this, a band of tin, *T*, being used to hold it down firmly upon the end of *B*. Screws, *S*, hold *T* down. A stove-bolt axle (See App. 93) is shown, and by using a nut, as explained, bolt magnets may be wound. By using the handle of App. 92, this arrangement can be used to wind almost anything, when used together with the attachment of App. 95.

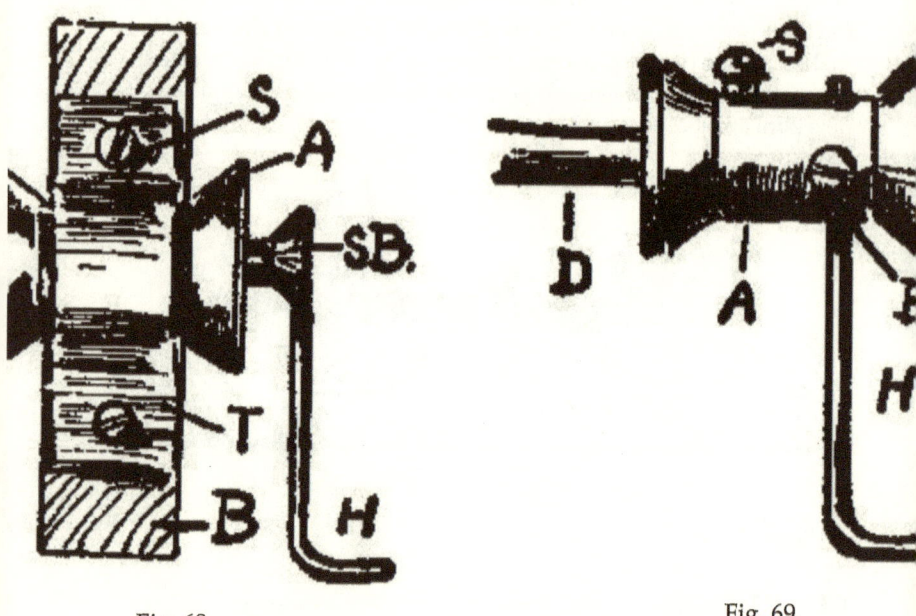

Fig. 68.   Fig. 69.

APPARATUS 92.

128. Crank for Winders, etc. Fig. 69. This form of crank or handle will be found easier to make than the one in which a wire is expanded in the slot of a stove bolt, and it can be used for many purposes, especially where dowels serve as axles. Wrap a little paper around the end of the ¼ in. dowel, *D*, and push it part way into [Pg 61] the spool, *A*, then put in a set-screw, *S*, to keep *A* from twisting upon *D*. The straight end of the wire, *H*, should be put into a hole, *B*, and another set-screw used to fasten it into the spool.

APPARATUS 93.

129. Winder. Fig. 70. For winding bolt magnets, this form of winder is very useful. It consists of a "stove bolt," *S B*, 2 in. long (total length) and 5/16 in. in diameter.

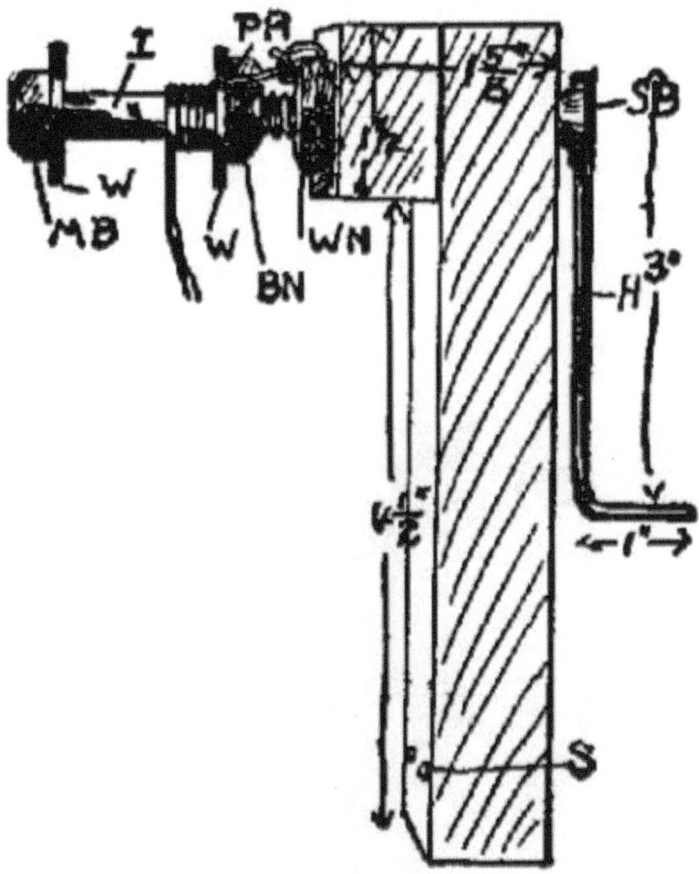

Fig. 70.

130. Handle or Crank, H, is made of a stout wire, 4 in. long, bent at the lower end as shown. H is fastened into the slot of S B. To do this the end of H is hammered flat until it will just slip into the slot. It may be soldered there, or be made to fit by expanding it so that it will press out against the sides of the slot. To do this, place S B into a hole in an anvil, or hold it in a vise, being careful not to injure the thread. Place the flattened end of H in the slot, and strike it on top so that it will expand and be pinched in the slot; but do not pound it so hard that you [Pg 62] split the bolt head. Three or four good center-punch dents upon the wire over the slot will help to expand it.

131. The Framework is made of wood, the dimensions being shown in Fig. 70. A 5/16 hole should be made for *S B*, the thread of which will stick through about ¼ in. so that the winder-nut, *W N*, can be turned onto it. *W N* should be on but 2 or 3 threads of *S B*. This will leave part of it for the thread of the bolt magnet, and when this and *S B* meet in center of *W N* they will bind against each other and hold the bolt tight. The winder can be nailed or screwed at S to the edge of a table or held in a vise.

APPARATUS 94.

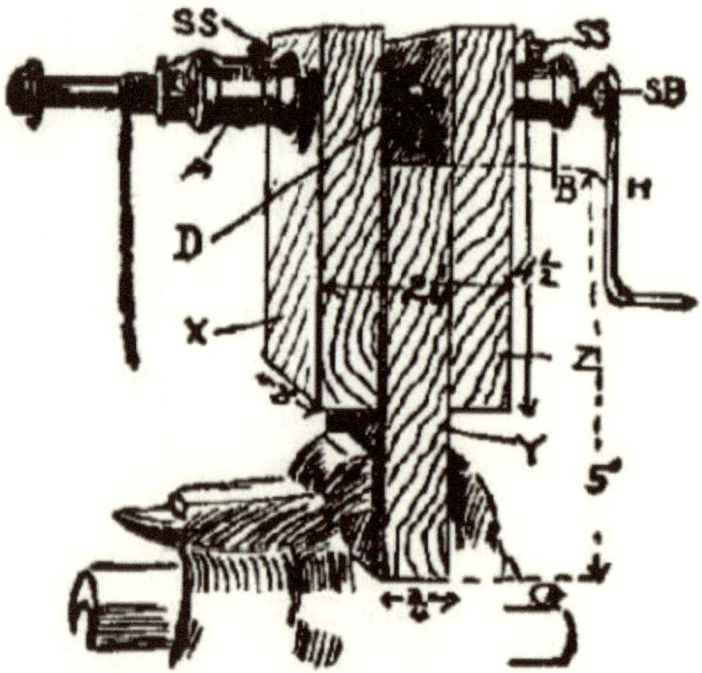

Fig. 71.

132. Winder. Fig. 71. This shows a winder that can be used for several purposes by arranging different attachments. It will be first described as shown in Fig. 71, where it is being used to wind a bolt magnet. The principal dimensions are shown in the figure. It is made of ¾ in. wood about 3 in. wide, the two outer parts X and Z being nailed to the center one, Y, which is to be held in a vise, or fastened to the edge of a table. A 5/16 in. hole should be made through the upper part X and Z at one side of the center, so that a long 5/16 bolt can be put through and used as described in App. 93, if desired. A smaller [Pg 63] hole, ¼ in., should be made on the other side of the center for a ¼ in. dowel. The dowel, D, is shown, and this size is a little smaller than the hole in ordinary spools, shown at A and B. One-quarter in. dowels can be made to fit fairly tight into the holes by wrapping paper around them. Five-sixteenth bolts can be screwed into the spool holes, shown by the bolt magnet in Fig. 71. To firmly hold a spool from twisting around upon the dowel-axle, a set-screw, S S, is needed. These are small screws, say ⅝ in. long, No. 5. A small hole should be made into the spool before forcing in the screw. (App. 25.)

The spools A and B are fastened in this way, by set-screws, to D. The handle, H, is made as in App. 93, in this case a *short* stove bolt, S B, being used and screwed into B. Fig. 69 shows a very simple form of handle for all such purposes, which may be used instead of the one here shown. The details of winding on the wire are given under App. 88.

APPARATUS 95.

133. Attachment for Winder. Fig. 72. By using this addition to App. 93 or 94, almost any ordinary kind of windings can be made. The wooden block, A, may be about 2 in. square and ⅞ in. thick. A set-screw, S, binds it to the dowel-axle, D, which is made to turn by one of the forms of cranks given, and which is held in one of the frameworks. Windings like that shown in App. 112, Fig. 85, can easily be done with this, the upright part, with the two spools, being screwed right to A of Fig. 72.

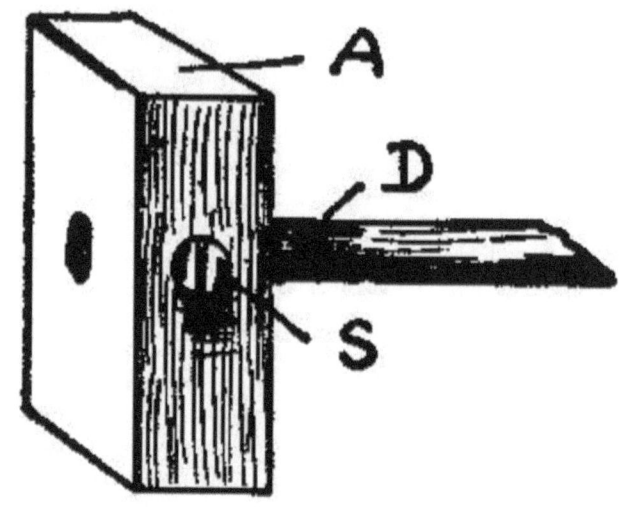

Fig. 72.

[Pg 64]

CHAPTER XI.

INDUCTION COILS AND THEIR ATTACHMENTS.

134. Induction Coils , or shocking coils, are rather expensive to buy, and altogether too complicated for boys to make by the methods usually given in books. The method here given is simple, the materials are cheap, and if you make them according to directions, you will have an apparatus that will, be able to make your friends dance to a rather lively tune. The amount of shock can be regulated perfectly (App. 103).

Winding. Full instructions have been given for making bolt magnets (App. 88). The winding of our induction coils is done in the same way by the same winder as the bolt magnets (App. 93), or by hand. You will find it a very tiresome and troublesome job, however, to wind on 12 or 15 hundred turns of fine wire by hand. Make a winder.

Several different forms of induction coils are shown. The *coil* is the most important feature, however, and we shall consider that separately. When you understand the construction of one coil, you can readily apply this to the different forms. Some form of *contact breaker*, or *current interrupter*, is needed also. These will be treated by themselves. The *connections* will be discussed under each form of apparatus.

APPARATUS 96.

135. Induction Coil; Construction of Coil Proper. Figs. 73, 74. An induction coil is a peculiar and wonderful apparatus. There are at least two coils [Pg 65] to each one. These are both wound upon the same core. They are made of different sizes of wire, are wound separately, and the strangest thing of all is, that these two coils are not connected with each other in any way. If they were not thoroughly insulated from each other, the coil would be of no value. (Study induction.) The winding of the two coils is done as explained in App. 88.

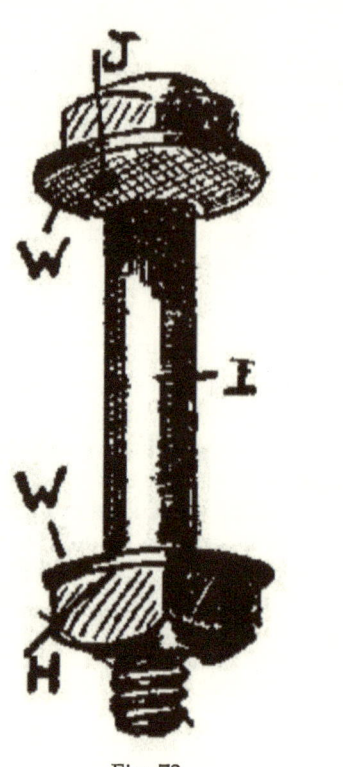
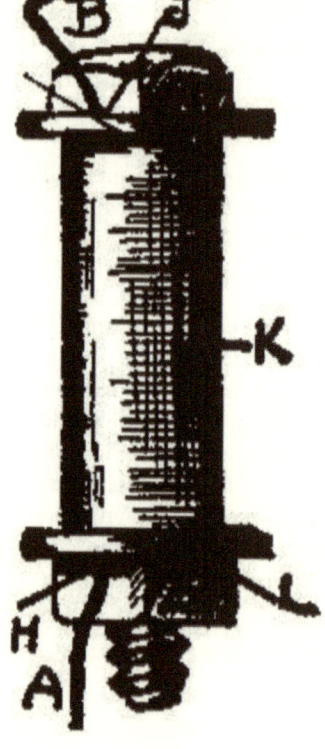

Fig. 73.

Fig. 74.

136. The Core is made of a 5/16 machine bolt, 2½ in. long. Leave but 2 or 3 threads at the end, just enough to fasten it solidly to the winder (App. 93). The washers should be about 1⅝ in. apart inside, and they should be made around a spool (§ 119) that is fully 1 in. in diameter.

137. The Inside or Primary Coil could be wound directly upon the bolt; but it is much better to cover the bolt with one or two thicknesses of paraffined paper, *I* (Index), as shown. A pinhole, *H*, in the washer is for the inside end (see § 123) of the primary coil, and the hole, *J*, is for the outside end of it.

The primary coil should be made of 3 layers of wire, which should be coarser than that used for the secondary coil. For our purposes it is best not to use a wire coarser than No. 20, and not finer than No. 24.

Use No. 24 insulated copper wire if you are going to connect ordinary batteries with it. A bichromate cell (App. 4) is best. Put about 6 in. (see § 109) of wire through H, and with App. 93 wind on 3 layers of say No. 24 wire. There being an odd number of layers, the winding will stop at the head end of the bolt, where a half [Pg 66] hitch (see § 110) should be taken before passing the wire through the hole, J. Cut the wire 6 in. from the hole. Write down the number of turns of wire to each layer and the total number of turns. You now have a 3-layer coil, and a current passed through this will magnetize the bolt; you have—so far—merely an electro-magnet. Cover the primary coil with 2 layers of paraffined paper, K (Fig. 74), and put some paraffine between the edges of K and the washers, so that the wire of the secondary coil cannot possibly come in contact with that already wound on.

138. The Secondary Coil should be made of a large number of turns of fine wire. Do not use anything coarser than No. 30. This is a good size, as finer wire is very easily broken by unskilled hands. For the size of bolt mentioned put on 13 layers. There will be about 100 turns to each layer, making a total of about 1,300 turns of No. 30 wire. Write down the total number of turns in *your* coil. To start the secondary coil, make a pinhole, L, just outside of the insulation, K, of the primary coil. Put 6 in. of wire through this, wind the end around the nut (App. 93, Fig. 70), and wind on as evenly as possible 13 layers. If the layers become rough, it is well to put a band of paper around after each 3 or 4. When you have finished take a half hitch (§ 110), and leave a 6-in. length free. Cover the secondary coil with strong paper. This coil may be used on any of the forms of shockers given.

APPARATUS 97.

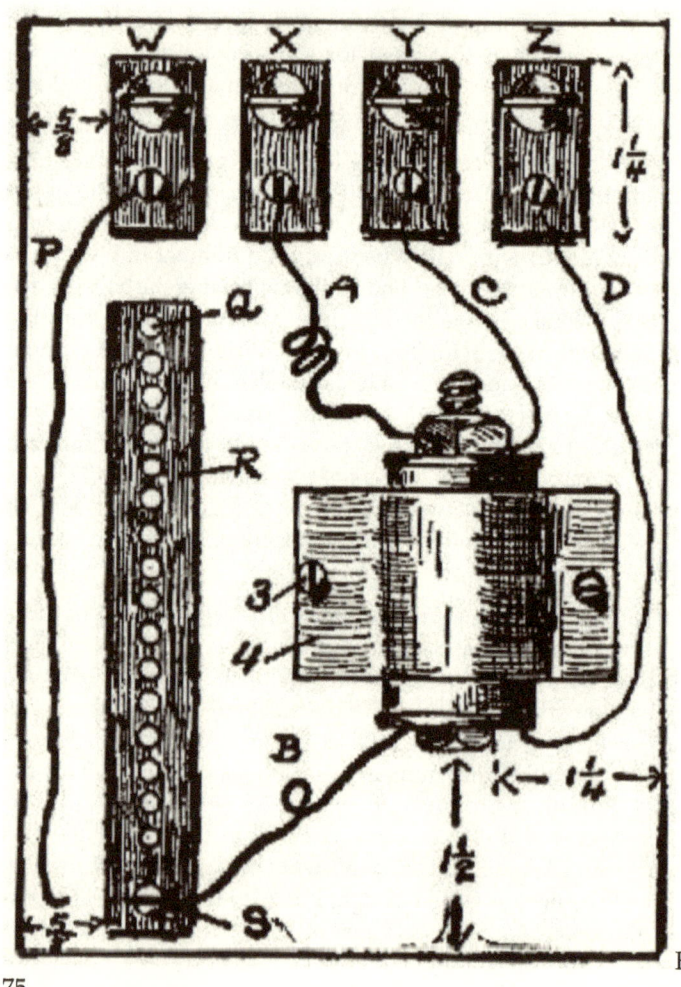

Fig. 75.

139. Induction Coil. Fig. 75. *The base* is made of a piece of board, 7 × 5 × ⅞ in. The locations of the different parts are shown in the figure. *The coil* is explained in detail in App. 96. It is fastened to the base by a thin copper strip, 4, which is bent over the coil and held down [Pg 67] by screws, 3. If you haven't any copper you can use a *narrow* strip of tin. Do not use a wide piece of tin or iron. The coil may be held down firmly by strong twine placed around each end of it. The

84

twine should pass through holes in the base, and be tied on the underside of the base. *The binding-posts* are like App. 46.

140. The Current Interrupter consists of a tin or copper strip, R, 6 in. long and ½ or ¾ in. wide. At one end of R is a screw, S, which is used as a binding-post for the outside end, B, of the primary coil. (See § 137.) Along the center line of the strip, R, are driven 1-in. wire nails, Q. These are placed ¼ in. apart, and they should go into the wood enough only to make them solid. (See Fig. 81.) Do not drive them in so far that they will split the base. A stout wire, P, fastened at one end only completes the interrupter.

141. The Connections. The binding-posts, W and X, should be connected with the wires leading from a battery. Use the bichromate batteries of App. 3 or 4. A dry battery will do. If the current enters at X, it will pass around the primary coil (§ 137) and out through B into R. It can go no farther until the free end of P is made to touch R, or one of the nails, Q, when the circuit will be closed. The current will fly around and around through the battery, primary coil, and interrupter as long as the end of P touches a nail. The battery current does [Pg 68] not get into the secondary coil at all. You can see, then, that the *primary circuit*, that is, the one passing through the coarse wire, will be rapidly opened and closed by bumping the free end of P along upon the row of nails.

The wires, C and D, coming from the secondary coil (§ 138) are in connection with Y and Z, to which are connected the wires leading from the handles (App. 101) held by the person receiving the shock.

142. To use the coil, arrange as explained. Let your friend hold the handles (App. 101) while you scrape the end of P back and forth along the row of nails. For those who cannot stand much of a shock, use a regulator (App. 103).

APPARATUS 98.

143. Induction Coil. Fig. 76. In case you wish to make the interrupter as a separate piece of apparatus, as App. 104, this arrangement will be found good. The *base* is 5 × 4 × ⅞ in. The *coil* is explained in App. 96, and the methods of holding it to the base are given in App. 97. The *binding-posts* are like App. 46.

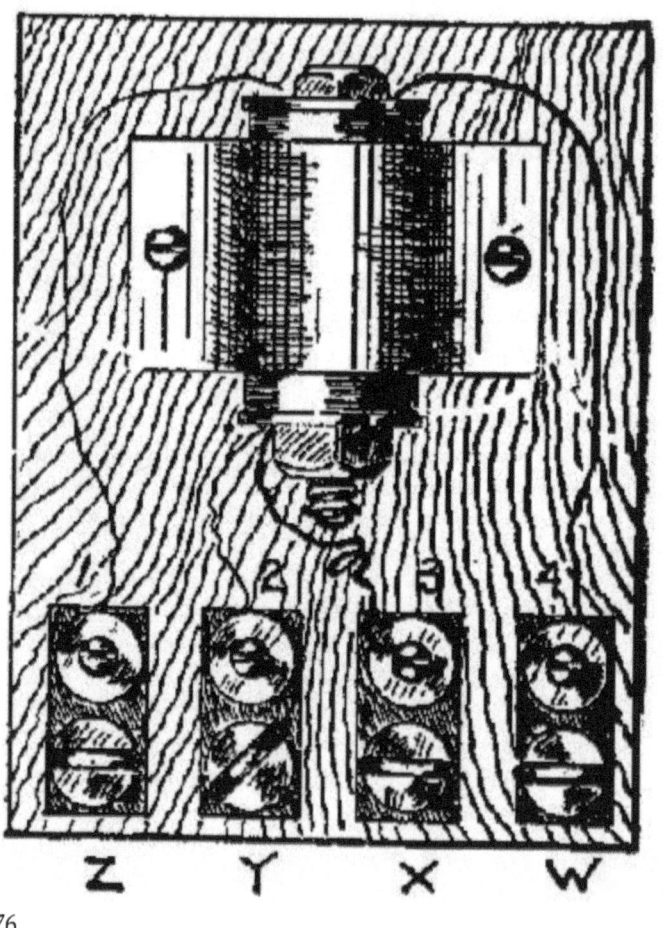

Fig. 76.

The Connections. We shall suppose that you have the interrupter of App. 104, Fig. 81. The ends of the primary coil (§ 137) are fastened under the screws of X and W, and those of the secondary coil to Y and Z. Connect one battery wire with X and the other battery wire to the interrupter at S, Fig. 81. Fasten the end of a stout wire to W, and leave the other end free to scrape along on the nails, Q, of the interrupter. This will then open and [Pg 69] close the primary circuit. The handles (App. 101) are connected with Y and Z, as explained in App. 97. Use the battery of App. 3 or 4.

APPARATUS 99.

144. Induction Coil. Fig. 77. If you wish to fasten your coil in an upright position the apparatus will look like Fig. 77. The *base* may be 5 × 4 × ⅞ in. The *binding-posts* are like App. 46. The *coil* is made as explained in App. 96; but to have all the ends of the coils come out at the bottom, as shown, an even number of layers of wire will be necessary. It will be just as well to have an odd number of layers as before, and to bring the wire ends down the side of the coil. The coil is fastened to the base with screws, S, passing through a tin strip, T, which has a hole punched for the bolt. T is squeezed between the regular nut on the bolt and an extra one on the underside of it. See Fig. 61 for suggestion of another method of holding bolts upright. *The connections* should be made with an outside interrupter, battery, and handles, as explained in App. 98.

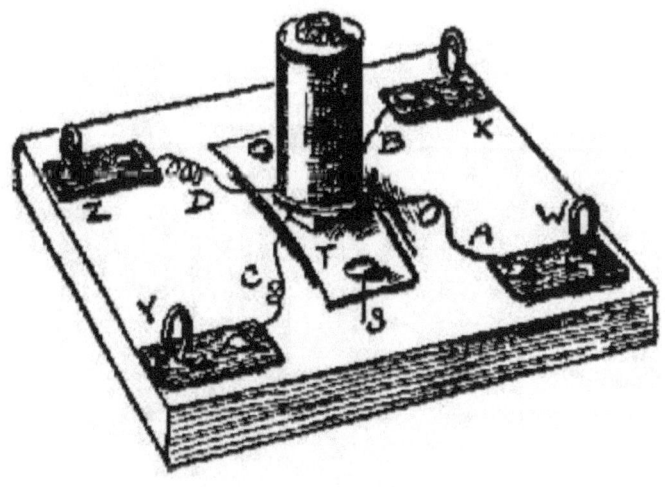

Fig. 77.

APPARATUS 100.

145. Induction Coil. Fig. 78, 78-A, 78-B. In case you wish to make a larger coil than those already described, the following will be found practical. It is made [Pg 70] in the same general way as before, an automatic interrupter, however, being added.

The Core is a machine-bolt, 4½ in. long and 5/16 in. in diameter. You may use a carriage-bolt of the same dimensions, if you file away the square shoulder at the head end, so that it will be the same size as the body of the bolt. Paste a piece of thick paper upon the head, so that A will strike the paper instead of the iron. *The Washers* should be made around a spool that is fully 1 in. in diameter. (See § 119.) The core should be insulated with paraffine paper before winding on the primary coil. (See App. 88.) The washers are 3⅞ in. apart, inside. *The winding of the coils* should be done with App. 93, or some other winder. The winder-nut, *W N*, Fig. 70, must hold the long core perfectly tight, to avoid wobbling. *The base* is 8 × 5 × ⅞ in. The different parts are placed as shown. The coil is fastened to the base as in App. 97. *For binding-posts* see App. 46.

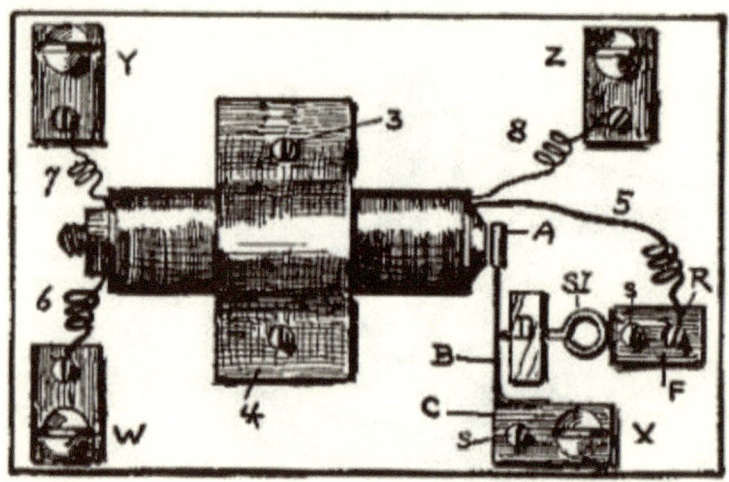

Fig. 78.

146. *The Primary Coil* (§ 137) is made by winding 3 layers of No. 24 insulated copper wire upon the insulated [Pg 71] core. One end, 6, is fastened to *W* (See § 109), and the other end, 5, is held under the screw-head, *R*. Wind at least two layers of paraffined paper around this coil before winding on the secondary coil.

147. The Secondary Coil (§ 138) is made of No. 30 insulated copper wire, there being 11 or 13 layers, each having about 200 turns. This makes, in all, about 2,500 turns of fine wire. If your winder works properly and the long core is strongly held by the winder-nut, you will have no trouble, although it takes a little time to wind on so many turns. The ends of this coil, 7 and 8, are fastened to Y and Z, which are made like App. 46. It will be found best to wrap a piece of thin paper around the coil after every 3 or 4 layers are wound on. This makes better insulation, and makes the winding easier. Protect the coil by covering it with thick paper. The whole coil, when completed, is about 1 in. in diameter.

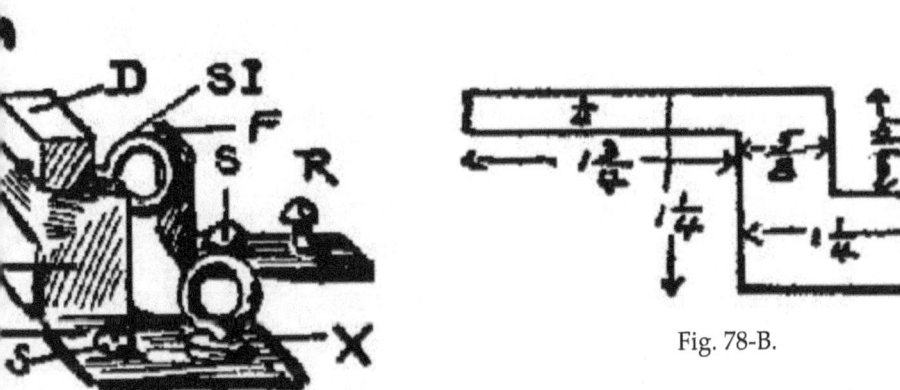

Fig. 78-A.

Fig. 78-B.

148. The Automatic Interrupter (Figs. 78, 78-A, 78-B) consists of several parts. B, E, C is a piece of *thin* tin, all in one piece. The part, B, is ¼ in. wide and 1¾ in. long. Its exact height above the base will depend upon the diameter of your coil. For the coil here described, 1 in. in diameter, the top edge of B is ⅝ in. above the base. See Fig. 78-B for shape of B, E, C before bending it, and for its dimensions. Around the end of B are tightly wound [Pg 72] several turns of tin, making the armature or hammer, A, which should not be allowed to strike against the head of the bolt on account of residual magnetism. (See text-book.) A piece of thick paper pasted on the head for A to strike upon is best. *A* will probably not get near enough to the bolt to strike it, but this will depend upon how you arrange the parts.

D is a wooden piece, 1 in. high, 1 in. wide, and ⅜ or ½ in. thick; it is nailed to the base. Through its center is a hole for the screw-eye, *S I*, which is the regulating-screw. *F* is a piece of copper, brass, or tin, ⅝ × 1¾ in. It is held to the base by the screw, *S*, and is bent so that it presses tightly against *S I*. Through *F* is a screw, *R*, to hold one end of the primary coil.

149. Adjustment and Use. The battery wires should be joined to *W* and *X*, and the handles to the secondary coil at *Y* and *Z*, unless a regulator (App. 103) is used. Let us consider the primary circuit. If the current enters at *W* it will pass through the primary coil and out at *X*, after going through 5, *R*, *F*, *S I*, *B*, *E*, and *C*. The instant that the current passes, the bolt becomes magnetized; this attracts *A*, which pulls *B* away from the end of *S I*, thus automatically opening the circuit. *B* at once springs back to its former position against *S I*, as *A* is no longer attracted; the circuit is closed and the operation is rapidly repeated. *B* should press gently against *S I*, which must be screwed back and forth, until the best results are obtained. While not in use *A* should be about ⅛ or 3⁄16 in. from the bolt-head. The armature, *A*, should vibrate back and forth very rapidly. If this coil gives too much shock with one cell of App. 3 or 4, put a regulator (App. 103) between *Y* and one of the handles (App. 101).

[Pg 73]

APPARATUS 101.

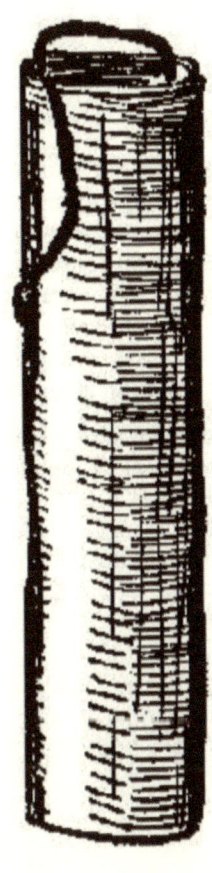

Fig. 79.

150. Handles for Shocking Coils. Fig. 79. Ordinary sheet-tin makes good handles. Cut 2 pieces, each 6 × 4½ in., and connect a stout copper wire to each. This may be done as suggested in Fig. 79, where the tin laps tightly over the bare end of the wire, or by punching 4 or 5 holes through the tin, and weaving the wire back and forth through the holes. Be sure that a tight and permanent connection is made. The wires joined to the handles should be about No. 20, and be 4 or 5 feet long. Roll the tin into a cylinder, so that the connection will be on the inside.

APPARATUS 102.

151. Handles for Shocking Coils. Very neat handles may be made from 4-in. lengths of brass tubing that is about ¾ in. in diameter. The wires leading to the coil may be soldered to the handles.

APPARATUS 103.

152. Current Regulator for Induction Coils. Fig. 80. If your coil gives too much of a shock with one cell of App. 3 or 4, you can pull the carbon and zinc partly out of the solution to weaken the shock, or you can use a water regulator. T is an ordinary tin tomato can nearly filled with water, L is a lamp chimney. One wire, A, is fastened to T directly, or by a spring binding-post. The other wire, B, is fastened to a piece of copper, C, which may be raised or lowered inside of L. D is a piece of pasteboard with a small hole in its center.

153. *Use.* If this apparatus be put anywhere in the primary circuit, the amount of shock can be regulated by [Pg 74] raising or lowering C. When C is raised, the current has to pass through a longer column of water than it does when C is near the bottom of L. When C touches T, the current passes easily. If it were not for the chimney, the current would pass to the sides of T.

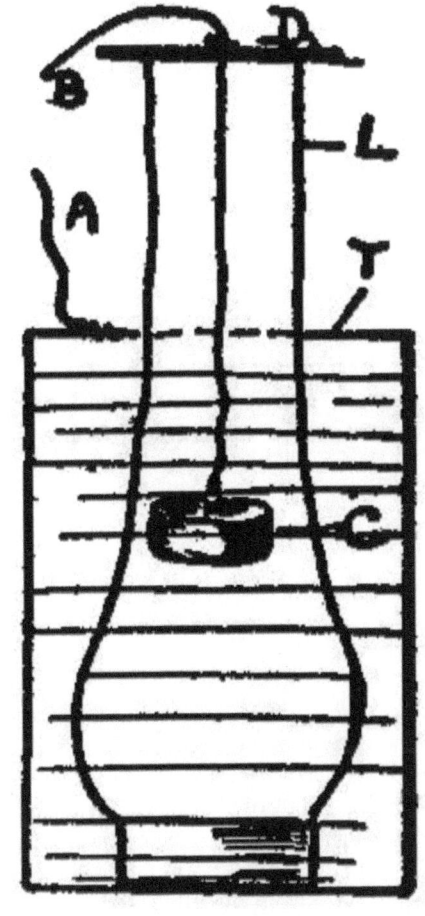

Fig. 80.

[Pg 75]

CHAPTER XII.

CONTACT BREAKERS AND CURRENT INTERRUPTERS.

154. Contact Breakers; Current Interrupters. It is often necessary to make and break the electric current at frequent intervals. This can be done by an ordinary key (App. 118) by rapidly raising and lowering it. It is more convenient, however, to use some other form of apparatus. The current may be interrupted automatically; that is, it may be made to do the work itself (App. 100), or each make and break in it may be governed by the student.

APPARATUS 104.

Fig. 81.

155. Interrupter. Fig. 81. The body of this consists of a strip of wood, 6 or 7 in. long, 1½ in. wide, and ⅞ in. thick. Cut a strip of tin 1 in. wide and long enough to bend down over the ends of the wood. Fasten the tin to the wood with small wire nails, driving the nails into the ends as well as into the top of the strip. Make a "center line" along the tin as a guide, and then drive 1-in. wire nails through the tin into the wood, so that they will make a row the length of the wood, and stand about ¼ in. apart. On one end make a hole through the tin, and put in a screw-eye binding-post (App. 45). It is evident that if a wire from one pole of [Pg 76] a battery be connected with the binding-post, it will also be electrically connected with the tin strip and nails. By touching the wire from the other battery-pole to the tin or to any nail, the circuit will be closed. If this last-mentioned wire be drawn along entirely above the tin, so that its end can bump along from one nail to another, you can see that the current will be closed every time a nail is touched, and be opened every time it jumps through the air. This apparatus can be connected with shocking coils, induction apparatus, etc., etc. Its use will be more clearly shown in connection with such apparatus.

APPARATUS 105.

Fig. 82.

156. Interrupter. Fig. 82. The nails in this apparatus are placed in a circle about 4 in. in diameter. They are electrically connected to each other by a bare copper wire, which is wound around each nail several times, and then led out to one of the binding-posts. In the center of the circle is a nail, or screw, which is connected by a wire to the other binding-post, care being taken not to allow the two wires to touch each other. Around the central screw is wound one end of a stout wire, the other end of which reaches out from the screw far enough to touch the nails. When this stout wire touches any nail, a current entering one binding-post can pass through nails, screw, etc., and out at the other binding-post. When the end of the stout wire is between two nails, the current cannot flow. By placing the finger against this stout wire and turning it around rapidly, the [Pg 77] current can be interrupted as desired. The base should be about $5 \times 6 \times 7/8$ in.

APPARATUS 106.

157. Interrupter. Wind the end of the wire from one pole of the battery around the handle of the file. Scrape the other wire along the rough file. As it jumps from one ridge to another the current will be rapidly interrupted.

APPARATUS 107.

158. Interrupter. Hold the end of the wire from one pole of a battery upon a saw-blade. Draw the other wire along over the teeth of the saw. As the wire jumps from one tooth to the next the current will be broken.

APPARATUS 108.

159. Automatic Interrupter. An ordinary electric bell, or buzzer, may be used as an interrupter. Every time the vibrating armature swings, the circuit is opened. The combination of a battery, induction coil, and electric bell makes a very good outfit for medical purposes. The automatic interrupter used on App. 100 should be studied.

[Pg 78]

CHAPTER XIII.

CURRENT DETECTORS AND GALVANOMETERS.

160. Current Detectors; Galvanometers. When a wire carrying a current of sufficient strength is properly brought near a magnetic needle, the latter will be deflected from its N and S line. The conducting wire has a magnetic field while the current passes through it, and this gives the wire the power to act upon a magnetic needle just as another magnet would.

The action of detectors, etc., depends upon this fact; and, strange to say, the magnetic field about the wire disappears the instant the current ceases to pass. The combination, thus, of a coil of wire and a magnetic needle, properly arranged, makes an instrument with which the presence of electricity can be detected. When the strength of a current is to be measured, or the strengths of two currents are to be compared, the apparatus is called a galvanometer. The method of making these pieces of apparatus will depend upon the strength of current to be tested or measured.

APPARATUS 109.

161. Current Detector. Figs. 38 and 40 show magnetic needles. These may be used to detect a current by holding the conducting wire near them and parallel to the needle. This form is not sensitive to weak currents. The delicacy of the apparatus is increased by allowing the wire to pass above and below the needle several times as in the next apparatus.

[Pg 79]

APPARATUS 110.

162. Current Detector. Fig. 83 consists, like all detectors, of a coil and a magnetic needle. The other parts are merely for convenience. Each turn of the coil helps to move the needle when the current passes.

Fig. 83.

163. The Coil is made by winding 10 feet of No. 30 insulated copper wire around the end of a broom-handle or other cylinder that is about 1 inch in diameter. This length of wire makes about 32 turns around such a cylinder. The exact length of wire for this makes no difference. After winding it, the coil should be slipped from the handle, being careful to hold it in such a way that it cannot uncoil and spring away from you. Tie the coil together with thread, in 3 or 4 places, to keep it in shape, and leave 5 or 6 in. of wire free at each end, so that connections can be made with other pieces of apparatus. After this is done press the coil into the shape shown, Fig. 83. This brings the wire near the needle and allows a longer needle to be used. The coil may be fastened to a pasteboard base. To do this, prick 4 holes in the base near the ends of the oval coil, and pass a strong thread through these with the aid of a sewing-needle. Tie the thread on the underside of the base at each end. If this is well done, the coil will be held firmly in an upright position. Paraffine may be used instead of the thread.

The ends of the wire should be made bare, and these may be sewed to the base to keep them in place.

164. The Needle may be supported upon a pin or needle-point. The piece of needle should be stuck through a cork which has a slot cut into its underside, so that it will straddle the lower part of the coil. The height of the [Pg 80] needle-point should be fixed so that the horizontal ends of the magnetic needle will be near the axis of the coil, that is, along its central line.

165. To Use the Detector, turn its base around until the coil is in the N and S line—that is, until the magnetic needle is parallel to the length of the coil and wholly inside of it. Touch the ends of the coil with the two ends of the wire, which is supposed to carry a current. The needle will fly around until it is nearly perpendicular to its former position, if the current is strong enough.

APPARATUS 111.

Fig. 84.

166. Current Detector. Fig. 84. To make a more substantial detector than App. 110, the coil should be fastened to a wooden base. The coil may be made of 10 ft. No. 30 wire, as explained. (§ 163.) A hole should be made in the base with a small awl or with a hot wire, and into this should be set a pin, head down. The hole need not be larger than the pin-head, and when you find out how high the pin-point should be above the base, the pin may be fastened in place with a little paraffine, which should be pressed into the hole around the pin. The coil may be fastened in place with paraffine. The ends of the coil may be connected with binding-posts, described in App. 46, as shown, or with any other desired form.

The base should be 4 × 5 × ⅞ inches. The coil looks well when placed about 1 in. from the edge of the base. The binding-posts may be about 1 in. from the edges.

[Pg 81]

APPARATUS 112.

167. Current Detector. Fig. 85. This is more troublesome to make than App. 111, but perhaps it looks more scientific.

Fig. 85.

168. The Coil is wound around 2 ordinary spools which are glued to a vertical piece, which, in turn, is screwed to a base. You should not use *iron* nails or screws in the construction of electrical apparatus, when a magnetic needle is to be used in connection with it, as these would attract the needle. The spools may be pushed onto dowels which are fastened into the vertical piece. Small brass screws are good for the purpose also, if you haven't good glue or the dowels. This coil, etc., may be used in connection with an astatic needle. The coil may be wound with App. 93 or 94, if you make the attachment of App. 95, and screw the upright carrying the spools to the attachment.

The *binding-posts*, shown in Fig. 85, are not to be advised. It will be better to use those of App. 45. The *magnetic needle* is supported by a sewing-needle stuck through a cork. This may be fastened to the base with paraffine.

169. It is often troublesome to turn the apparatus around until the needle becomes parallel to the length of the coil. To avoid this, a small bar magnet, shown in the Fig. 85, may be laid on top of the coil. A magnetized sewing-needle will do, and this will keep the magnetic needle quiet and parallel to it when the current is not passing through the coil. Of course, it takes a little more current to move the magnetic needle when the bar magnet is in place, than it does without the magnet.

170. By allowing the current to enter the right-hand [Pg 82] binding-post, as you look at it from the front (Fig. 85), it will go around the coil in the direction of the hands of a clock, that is, from left to right on top. This, of course, is not necessary to merely detect the presence of a current. In order, however, to determine the direction of currents by means of a magnetic needle, study the effect with a single turn of wire at first. (See text-book.)

171. *Dimensions.* The base is 5 × 4 × ⅝ in. The upright piece is 5 × 3½ × ⅝ in. The spools are 2½ in. apart center to center.

APPARATUS 113.

Fig. 86.

172. *Astatic Current Detector.* Fig. 86. The ordinary magnetic needle points to the north quite strongly. It is evident, then, that this pointing-power must be overcome by the magnetic field around the coil of wire, before the needle can be forced from the N and S line. *Very* weak currents will not visibly move the magnetic [Pg 83] needle in the detectors so far described. You should remember that no action will take place unless the magnetic field around the magnetic needle is acted upon by that around the coil. In order to make an instrument that will be very sensitive, we must have strong fields about the needle and coil, and we must, at the same time, decrease the pointing-power of the needle. We can increase the strength of the field about the needle, and at the same time decrease its pointing-power by using an astatic needle. (See App. 69.) The arrangement shown in Fig. 86 is a very simple one, and it is quite sensitive.

173. Details of Construction. The base is 4 × 5 × ⅞ in. The *coil* is made from 10 ft. of No. 30 insulated copper wire. (See § 163 for details about coil making.) The *binding-posts* are like App. 41. The *Astatic Needle* is described for App. 69. The needles may be broken off, if too long for the coil. They are supported by a fine thread hung from a screw-eye, which may be turned to adjust the position of the needles. This is not necessary, as the thread may be hung from a plain wire arm that reaches out from the upright rod. This rod is a 6-in. piece of dowel, ¼ or 5/16 in. in diameter. It stands in an ordinary spool which should be glued to the base. Do not nail it to the base. The wire arm may be of iron, as it is some distance above the needle; but it is better to use a stiff brass or copper one. In the figure one end of the wire is twisted around the screw-eye, making a nut for the screw-eye to turn in.

Hang the astatic needle so that the wire between the two parts will not quite touch the coil. The needles should be parallel to the coil before testing for currents. They will fly around very decidedly with even fairly weak currents.

[Pg 84]

APPARATUS 114.

Fig. 87.

174. *Astatic Current Detector.* Fig. 87. For a description of the wood-work, coil, etc., see App. 112; for the astatic needle see App. 69; for the method of supporting the needle see App. 113, Fig. 86. The top part of the coil is spread apart a little to allow the lower needle to be dropped through the opening thus made, and to allow the wire joining the two needles to be free to turn. The needles may be broken off a little, if necessary, or an opening may be cut into the vertical part of the frame, so that they can swing more freely. This detector will indicate quite feeble currents.

APPARATUS 115.

175. *Astatic Detector.* Fig. 88. As previously stated, the sensitiveness of a detector can be made greater by increasing the strength of the coil-field for a given current. This may be done by increasing the number of turns of wire in the coil. The most convenient way will be to use two coils, one on each side of the astatic needle.

176. The Support, or framework, is a lamp chimney. By this the astatic needle is suspended and protected from air currents. The chimney should be at least 3 in. in diameter at the bottom, about 10 in. high, with a plain round top. Upon the top of the chimney is placed the cover of a wooden pill-box, 2 in. in diameter.

177. The Coils should be made separately, for convenience. Each should be of 10 ft. No. 30 wire. (See details § 163.) Cut out a round piece of stiff pasteboard, just large enough to go inside of the bottom of the chimney. Fasten the coils to this by sewing (§ 163), or with [Pg 85] paraffine, so that they shall be symmetrically located and ⅜ in. apart. The pasteboard circle may be fastened to the base with small brass screws. Do not use any iron nails or tacks. In this, all four ends of wire are brought out under the edge of the chimney (Fig. 88). Cut little grooves in the base for the wire to sink into, so that the chimney will rest firmly upon the base all around. The ends of the wires are fastened to three binding-posts.

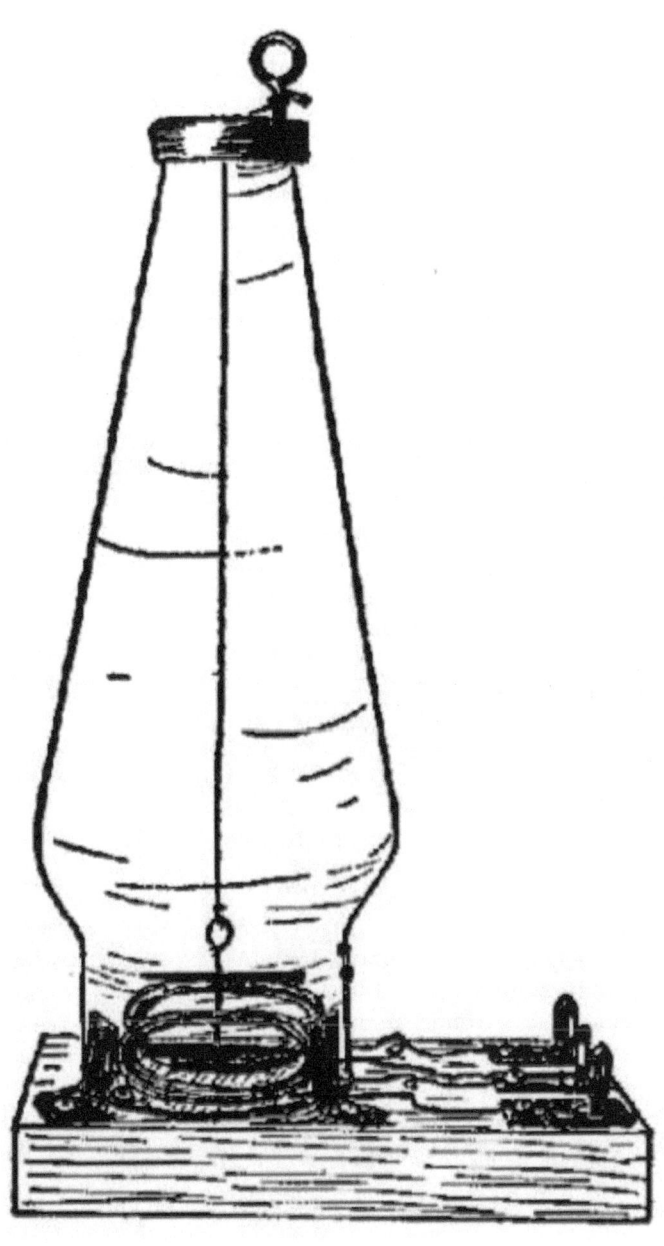

Fig. 105

88.

178. Joining the Coils. The end of one coil must be joined to the beginning of the other properly, or the action of one will destroy that of the other. Fig. 89 shows the two coils, *A* and *B*. If the current enters at the binding-post, *X*, it will pass through the turns of coil *A*, in the direction of clock-hands, then out to *Y*, where *B* begins, around *B* in the same way, and then to *Z*. *Y* may be [Pg 86] simply a screw-eye binding-post (App. 41). By this arrangement one or both coils can be used at a time. If the current is very weak, use both coils; that is, connect the ends of wires to be tested with the two outside binding-posts. If they are joined to the middle and one outside post, one coil only will be in the circuit.

179. The Base should be about 7 × 5 × ⅞ in. Fasten three bent brass or copper strips to the base with brass screws to hold the chimney steady. By bending them in more or less you can make a snug fit around the chimney.

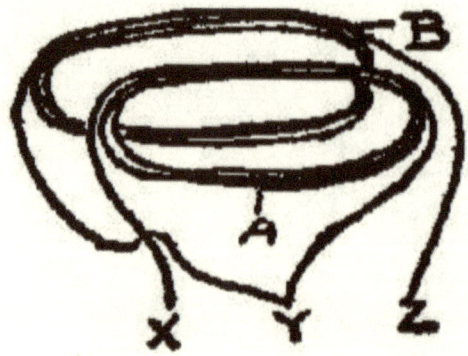

Fig. 89.

180. Adjusting the Needle. In the center of the box-cover is a small hole. The thread from the needle passes through this. The upper end of the thread is wound around a screw-eye, which is screwed into the cover near one edge. By turning the cover around, the needle can be made to hang parallel to the coils, and by turning the screw-eye, the needle can be raised or lowered. A small hole should be made in the cover before putting in the screw-eye, or you will be liable to split the wood.

181. Use. This apparatus will indicate very slight currents; in fact, as feeble ones as the student will have occasion to experiment with, such as induced currents, currents of thermo-electricity, and currents produced by exceedingly weak batteries. (See text-book.)

APPARATUS 116.

182. Tangent Galvanometer. Fig. 90. For the uses of this form of galvanometer see text-book. Do not use any iron in making this apparatus. *The base* is 5 × 4 × ⅞ in. At its front end are three binding-posts. [Pg 87] *The pasteboard band*, G, is 1¼ in. wide and 6 in. in diameter. Cut the pasteboard 21 in. long and 1¼ in. wide, then bend it into the form of a circle. There will be a lap of about 3 in., and you can make it solid by sewing the two ends together at the lap.

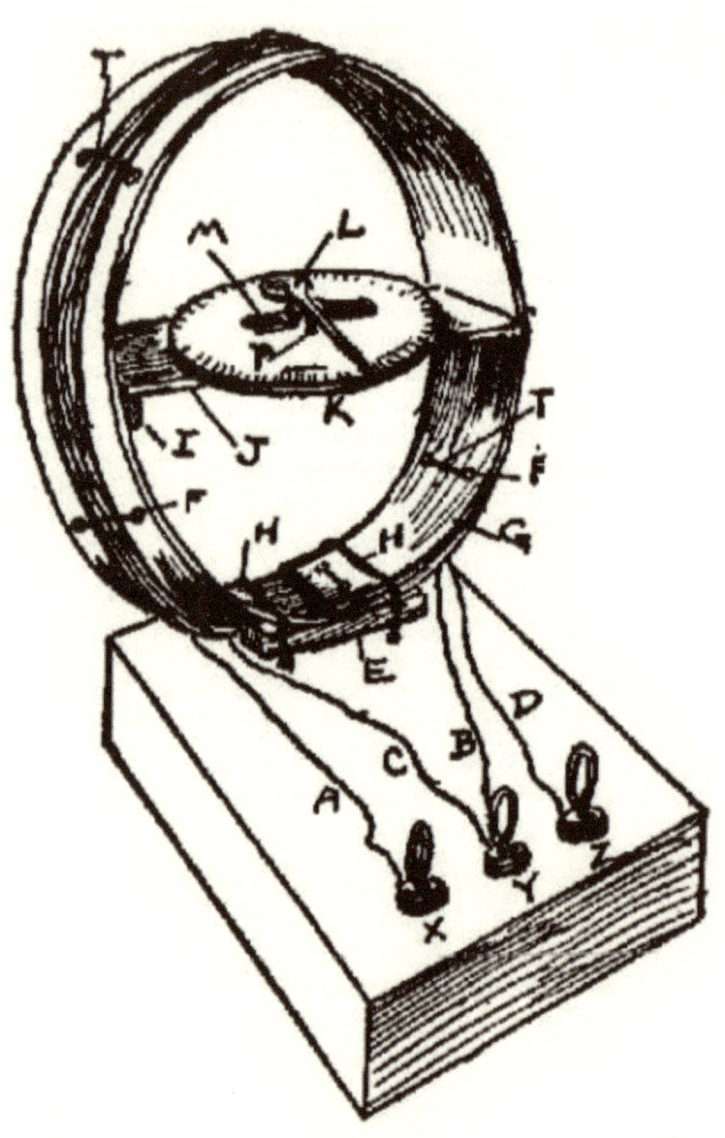

Fig. 90.

183. The Coils maybe made of No. 24 insulated copper wire, which should be wound on before fastening G to the base. There are two separate coils, one having five turns and the other ten turns. Leaving a 6-in. length, A, for connections, wind five turns of wire on to G, putting them on clockwise; that is, pass them over the top of G from left to right. Tie thread around G and the wire to hold them together after you have five turns on, and cut a 6-in. end, B. Now begin with C, and wind on ten turns, bringing the end of them out at D. Punch holes, F, through G on each side of the coils, run twine, T, [Pg 88] through them, and tie T on the outside of G. Do this in three or four places, to firmly hold the coils.

184. Fastening Coils to Base. The band and coils will not rest squarely upon the base, so cut two pieces of wood, E, about 2 × ¼ × ¼ in., to be put under G, one being on each side of the coil. Make holes through the base, pass strong cord, H, through them, and over the inside of G, then tie under the base. This should tightly squeeze E, and hold G upright and firm.

185. The Connections. A and B are the ends of the five-turn coil; C and D are the ends of the ten-turn coil. If the battery-wires are connected with X and Y, the current will pass through five turns of wire; if connected with Y and Z, it will pass through ten turns; if with X and Z, the current will pass through the entire fifteen turns. In this way the strength of the magnetic field about the coil can be regulated, and its effect upon the magnetic needle, M, changed.

186. To Support the Needle, glue or sew two strips, I, to G. They must be in such a position that the poles of M will be as nearly as possible in a horizontal line drawn through the center of the circle, G. After you have made M (App. 66), and have found where the pieces, I, should be, fasten them to G, and then to I glue a pasteboard strip, J, about 1¼ in. wide. Run a pin, P, up through the center of J to support M.

187. The Magnetic Needle, M, should not be over 1 in. long for this kind of an instrument. (See App. 66 for full directions for making it.) On the top of M should be fastened a light paper pointer or index, L. The short end should be made large, so that the long slim end will not over-turn M; that is, the pointer should balance itself. It may be fastened to M with paraffine or a drop of sealing-wax. If carefully balanced, the pointer can be made quite long.

[Pg 89]

188. The Graduated Circle, K, is described. (Index.) With this you can tell through how many degrees the needle is deflected, when the current passes. The strength of different currents can be compared, and many interesting experiments performed with the tangent galvanometer. For clearness, the circle, K, is shown small. In order to have the divisions on it far enough apart, K should be about 4 in. in diameter. The zero points should be at the front and back of the instrument, when a pointer is used on the needle.

189. How to Use It. For full explanations, and for the study of experimental cells, etc., by means of the tangent galvanometer, see text-book. It will be impossible for you to get M exactly in the center of G; you cannot get the pointer exactly at right angles with M; hence, if you pass a certain current through the coils, and the pointer reads 20 degrees, you will find, if you reverse the current, making it go through the coil in an opposite direction, that the pointer may read 24 degrees on the opposite side of the zero. To get the *true* reading, then, take the average of the two, which in the case mentioned would be 22 degrees. (See current reversers.)

APPARATUS 117.

190. Tangent Galvanometer. Fig. 91. *The base* consists of 2 parts, A and B. It is not necessary to use two pieces if you have wood that is at least ⅞ in. thick. This is given as a suggestion in case you have nothing but thin boards. By screwing B to A the base is made thick enough to take the screws for binding-posts. The base proper, A, is 8½ × 5 × ½ in. If you make this of ⅞ in. stuff, you will not need B.

The Back, C, is 10 × 8½ × ½ in. It is screwed to the base. Do not use nails, as these affect the magnetic [Pg 90] needle. Find the center of C, and with this as a center, draw two circles, (that is, the circumferences of two circles,) one 5 in. in diameter to show where to cut out a hole, H, and the other 7 in. in diameter to serve as a guide for fastening on the spools, F.

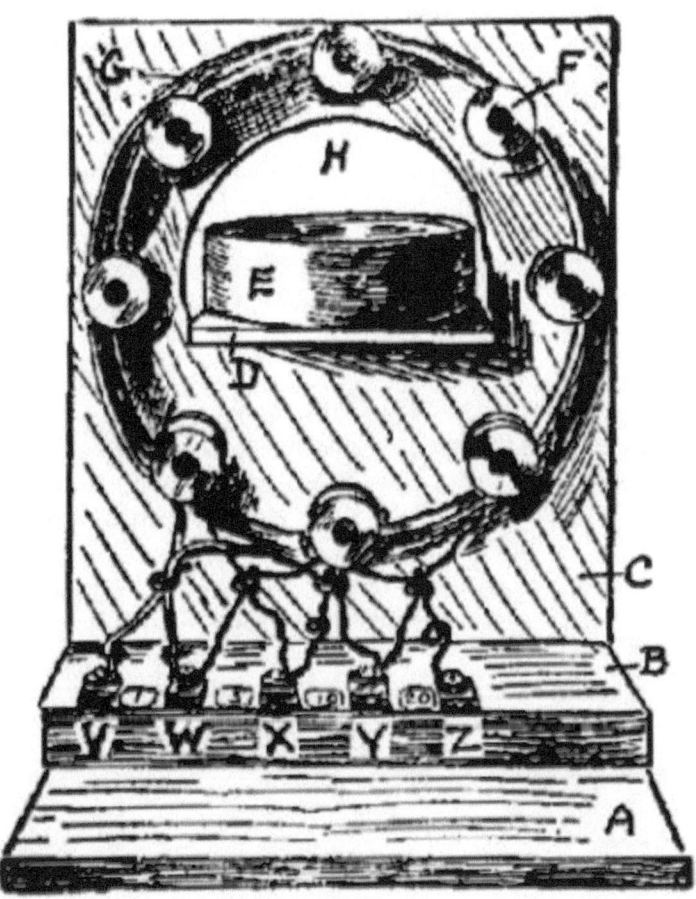

Fig. 91.

The Spools, F, are glued to C. If you have brass screws, these may be used instead of the spools; they should be left sticking out from C about 1 in. Around the spools or screws, fasten a pasteboard band, G, on which to wind the wire. G may be about 1 in. wide; it should be kept in the circular form by sewing the ends together where they lap. (Read directions in App. 116.)

191. The Coils on this model are 4 in number. (See App. 116 for the method of winding.) The first coil is made of coarse wire, No. 18, its ends being joined to the binding-posts, V and W. The second coil has 5 turns of No. 24 insulated copper wire, its ends being joined to W and X. The third coil has 10 turns of the same size wire, No. 24, and is joined to X and Y. The fourth coil has 20 turns of the same joined to Y and Z. If you want to use [Pg 91] the galvanometer for quite weak currents, it would be well to make a fifth coil of 20 turns of No. 30 wire, and join it with Z and a new binding-post. The ends of the coils are run through small screw-eyes before passing to X, Y, etc. This is not necessary, it merely keeps them in place.

The Binding-Posts are like App. 43. Any other desired style may be used, those of App. 46 being preferred.

The Hole, H, is 5 in. in diameter. It should be cut out about ½ in. below the center of the circles to allow for D, and for the pin-point which supports the magnetic needle, the poles of which should be in the line passing through the center of the coils. The method of cutting the hole, H, through C, will depend upon the tools at your service.

D is the front edge of an adjustable table, like that explained. (Index.) It is 4¼ in. wide. It supports the magnetic needle which is inside of E.

E is the outside of a glass-covered compass. (See App. 67 for details.) The needle should not be over 1 in. long.

[Pg 92]

CHAPTER XIV.

TELEGRAPH KEYS AND SOUNDERS.

APPARATUS 118.

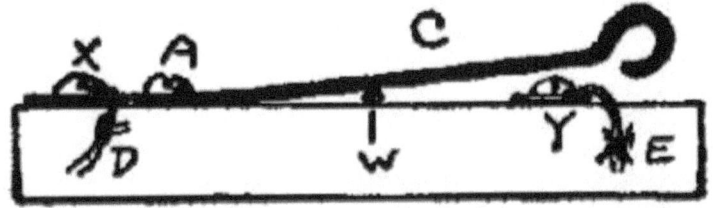

Fig. 92.

192. Telegraph Keys. Fig. 92. Telegraph keys are merely pieces of apparatus by which the circuit can be conveniently and rapidly opened or closed at the will of the operator. An ordinary push-button may be used to turn off and on the current, but it is not so convenient as a "key." Fig. 92 shows a side view of a simple key. C is a metal strip about ¾ in. wide and 4 or 5 in. long. At the left end it is fastened to the base with a screw, A. Another screw, X, serves as one binding-post. Y is another screw binding-post. W is a short wire, used to regulate the amount of spring to the key. This is done by moving W to the right or left. If the current enters at X, it will pass along C and out at Y, when C is pressed down. By moving C up and down according to a previously arranged set of signals, messages can be sent by means of the electric current. (See telegraph alphabet.) This apparatus is not a good one where the line is to be run with a "closed circuit battery," or where it is to be used very often. It will do, however, for places where a push-button would be too tiresome to use. The right end of C is curved. This curve serves as a handle. D and E are wires leading from X and Y.

[Pg 93]

APPARATUS 119.

193. Telegraph Key. Fig. 93. *The base* is 5 × 4 × ⅞ in. *The key*, C, is made of two thicknesses of tin. It is made into a strip 5½ × ¾ in., then the front end is bent up for a handle, as suggested in Fig. 92, the front end being above the base so that it will not touch the strap, D, unless it is pressed down. C is fastened to the base by a screw, H, which also binds one end of the copper wire, C W. About ¾ in. from H is placed X, which is a screw-eye binding-post. Under C is the wire, W, which is used to regulate the amount of spring in C, by moving it forward or backward. S I shows the position of a screw-eye, or of an ordinary screw put into the base through C. The hole in C should be made so that C can move up and down easily around the screw. This is used to make a click when the key is allowed to spring up. The downward click is made when C strikes D at each depression.

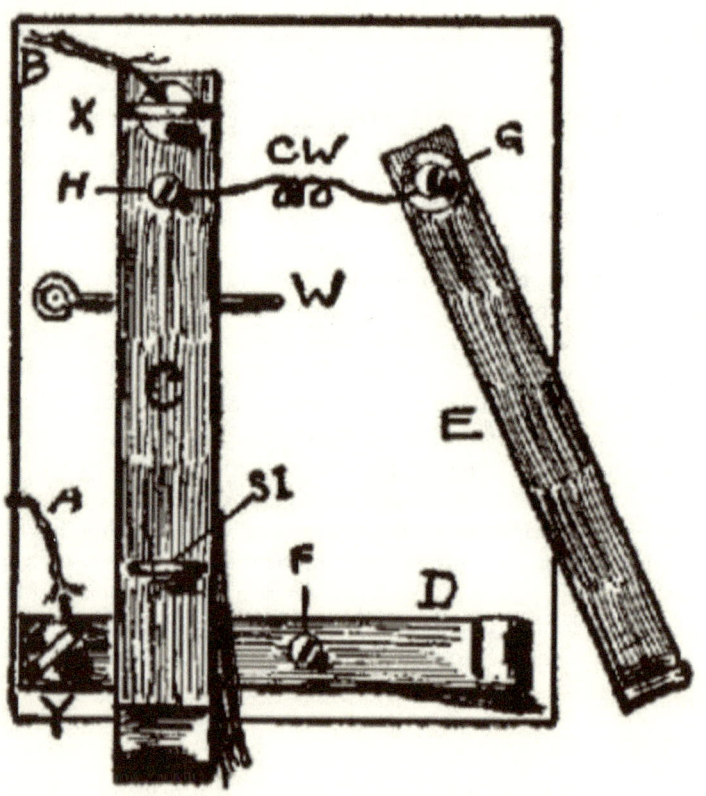

Fig. 93.

The Strap, D, is made of tin. It is 4 × ½ in. before bending up the right end a little. It is fastened to the base by the screw, F, and by the other binding-post, Y. [Pg 94] Its right end is raised enough to allow the arm, E, to pass under it, but it must press down well upon E when E is forced toward F.

The Swinging Arm or Switch, E, is also made of tin, and measures, finished, 4½ × ½ in. Its front end should be bent up a little for convenience in handling it. (See Fig. 92.) E is pivoted at G by a screw, which also binds the wire, C W. Fig. 24 shows another way to make the pivot and connection.

194. Operation. See Fig. 99 for the details of the connections of a home-made telegraph line. When you are using the line and telegraphing to your friend, the switch, E, of your instrument must be open, as in Fig. 93, and the corresponding switch on his instrument must be closed; that is, the circuit must be opened and closed at but one place at a time. As soon as you have finished, your switch must be closed. He will open his and proceed. When you have both finished, both switches must be closed. If your friend left his switch open, you could not call him over the line, as no current could pass into his sounder.

195. Batteries. As the circuit has to be left closed for hours and perhaps days at a time, so that either operator can call the other, a closed-circuit battery is necessary. (See App. 9.) A dry cell, Leclanché, or other open-circuit cell would not be at all suitable for a telegraph line, as it would soon polarize. Large Daniel cells, which are 2-fluid cells like App. 7, or gravity cells (App. 9) are the best for your line.

APPARATUS 120.

196. Telegraph Sounder. Fig. 94. The wood-work consists of 2 parts; the base, B, is 6 × 4 × ¾ in., and the back, A, is 6 × 5 × ½ in. A is nailed or screwed to B.

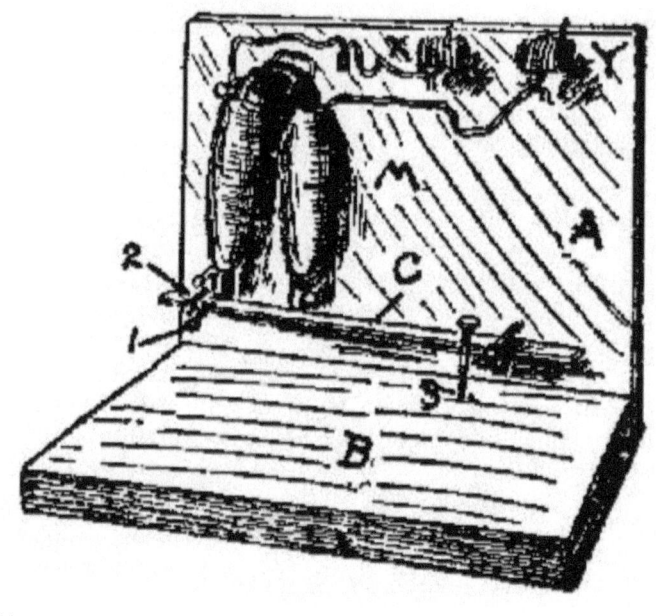

Fig. 94.

[Pg 95]

The Magnet, M, is fully described in App. 85. M is held firmly to A by cord or wire, which should pass around it near the poles and at the curved part. The wire should pass through small holes in A, and be tied at the back. Wire nails driven into A at the sides of M will keep it from moving about. The wires from the magnet coils are led to two spring binding-posts, X and Y.

197. The Armature, C, is made of a narrow piece of thin *iron*, about $5\frac{1}{2} \times \frac{1}{4} \times \frac{1}{8}$ in. It may be made by bending up 3 or 4 thicknesses of tin into that shape. This is the part which will be attracted by M, when the current passes, and which will make the clicks by which the message can be read. (See telegraph alphabet.) There are many ways by which C can be held near M. The figure shows how it can be done entirely with 1-in. wire nails. At the right end of C two nails are driven into A above and below C. They are just far enough apart to allow the left end of C to be raised and lowered without binding; in other words, these nails make a pivot for C to swing upon, and

they help to support it at the same time. The left end of C must not quite touch the poles of M when the current passes, because the residual magnetism would keep C from dropping back into place. To adjust the armature, pass the current through M, hold C so that it will not *quite* touch the poles, then drive in the upper nail, 2. Put another nail, 1, below C, so that M will not have to lift C more than ⅛ or 3⁄16 in. Try the nails in different positions until C quickly rises and falls when the circuit is closed and opened. A nail, 3, [Pg 96] driven in front of C, will keep its right end in place. No springs are needed, as gravity acts upon C instantly, bringing it to the lowest position as soon as the current ceases to flow.

198. The Battery will depend upon how much you want to use the sounder. If just to show the principle of it, almost any cell of medium strength will do, like that of App. 3, 4 or 5. A dry battery will do, but if you use the sounder much, an open-circuit battery will soon use itself up. Where much work is needed of the battery use App. 9.

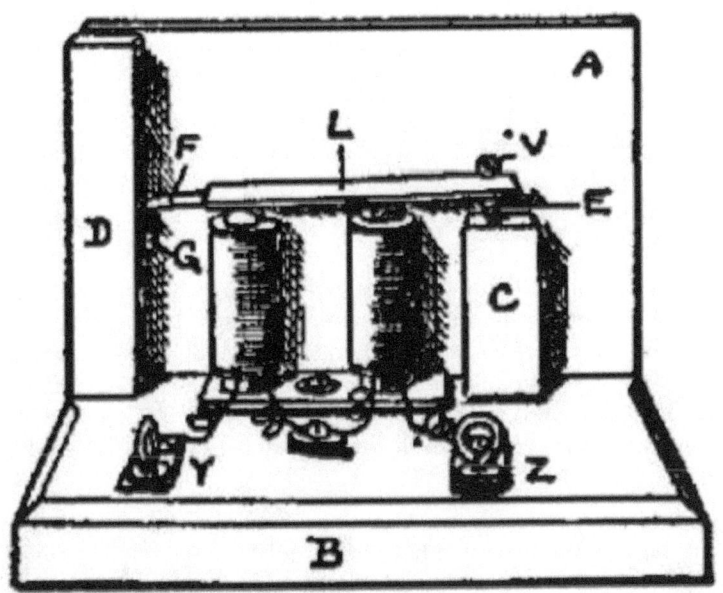

Fig. 95.

The Key like App. 119 is best. Push-buttons are handy where used only for experiments, and not for the actual sending of messages.

APPARATUS 121.

199. Telegraph Sounder. Fig. 95. This makes a simple and efficient sounder for short lines. The *base*, *B*, is 7 × 4½ × ⅞ in. The *back*, *A*, is 7 × 4½ × ½ in.; it is nailed to *B*. The piece *D* is 4 × ¾ × ¾ in.; it is nailed to *A*. *C* is a wooden piece 1½ × ¾ × ¾ in.; it is nailed to *A*, and in its top is a screw, *E*, which is used as a regulating-screw to keep the armature, *L*, from touching the poles.

[Pg 97]

200. The Armature, L, is explained as App. 77. The two thicknesses of tin at *F* must not be too thick, or it will take too much battery power to work the sounder. If you find that it is too stiff to bend down, when the current is on, try the arrangement of App. 122, which is easier to make and regulate. The whole point depends upon the tin you have. The end of *L* must tap against *E*. A hole is punched in the part *F*, and a screw, *G*, holds it to *D*. *L* should rest about ⅛ in. above the poles and gently press against a screw or nail, *V*.

201. The Magnets are like App. 89. They are made as in App. 88, and held down like App. 90. These should be placed very near the back, *A*, so that the armature will be over them. If your yoke is not too wide the coils may rest against *A*. *Y* and *Z* are binding-posts like App. 46.

202. Connections. Join the coils as explained in § 125 and see § 115. Instead of a third or middle binding-post, as in Fig. 66, hold the two inside ends between a screw-head and a copper bur. The method of joining the wires for a line with two outfits, is shown in App. 124. If you have but one key, sounder, and battery, simply join the line wire to the return wire there shown. A gravity cell is best. (See App. 9.)

203. Hints About Adjusting. If you have the right spring to the part *F*, of the armature, you will have no trouble. It must not be so weak that it allows *L* to strike upon the poles, as the residual magnetism (Text-book) will hold *L* down after the current has ceased to pass. No springs are necessary, if your tin is right. Do not have *L* too far away from the poles. The distance is regulated by the position of *V*. If you have trouble in getting it to work see App. 122. The poles must be opposite in nature.

[Pg 98]

APPARATUS 122.

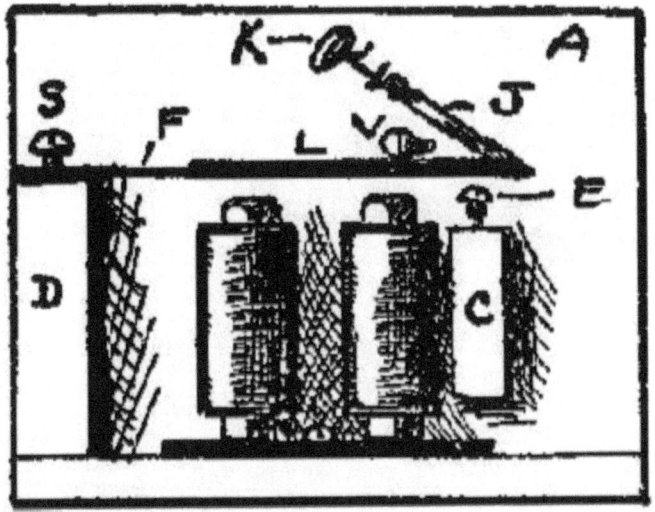

Fig. 96.

204. Telegraph Sounder. Fig. 96. The magnets, connections, etc., are like those of App. 121, no binding-posts, etc., being here shown. The armature is straight, however, the part *F* resting upon *D*. A hole is made in the end of *F*, and through this is a screw or nail, *S*. The hole must be large enough to allow *S* to pass through easily. This acts as a bearing or pivot. *L* is kept up against *V* by the rubber-band, *J*, one end of which passes around the end of *L*; to the other end of *J* is a thread, which is tied around a screw-eye, *K*. By turning the screw-eye, the band may be made to pull more or less upon *L*. In this way the apparatus may be regulated according to your battery. The general dimensions and explanations are given in App. 121. *D* is made of such a height that it will bring *L* about ⅛ or 3/16 in. above the poles.

APPARATUS 123.

Fig. 97.

205. Telegraph Sounder. Figs. 97 and 98. This apparatus looks a little more like a regular sounder than App. 121 and 122, but it is much harder to make and adjust. In this the lower nuts of the bolts are not sunk into the base, and the magnets are made of 2-in. bolts. If you change this and fasten them like App. 89 and 90, it will simply [Pg 99] change the dimensions of the small parts. The sizes given are for this particular instrument.

Fig. 97 shows a perspective view, and Fig. 98 is a plan or top-view of it, with dimensions.

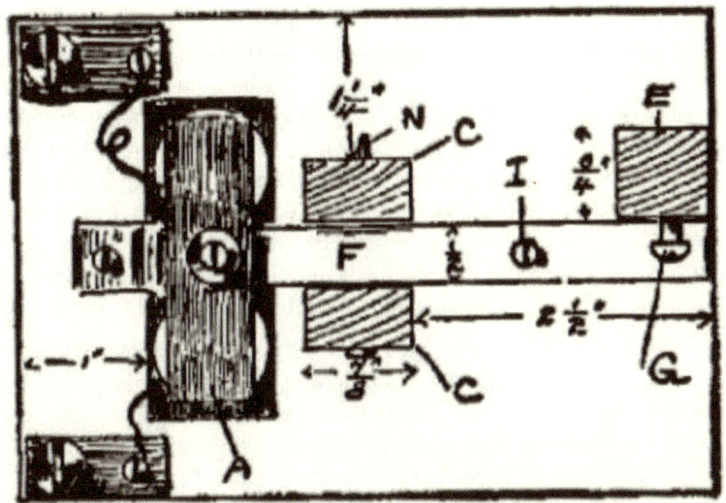

Fig. 98.

206. *The Base, B,* is 6 × 4 × ⅞ in. *The magnet, M,* is explained in App. 89. Its wires are attached to the binding-posts like App. 46. *The armature, A,* is 2½ × ¾ × ⅛ in., and made as described in App. 71. *The piece, D,* is 2½ × 1⅜ × ½ in., and is screwed to B from below, after the two uprights, C, are nailed to it. *The uprights, C,* are 2¾ × ⅞ × ½ in. They are nailed to D. *The nail, N,* runs through both uprights, and acts as the bearing for F to rock up and down upon. The hole for N is 2 in. above B. It must not be too loose in the holes, or F will rock sidewise, and allow A to touch one of the magnets. *The upright, E,* is 2¾ × ¾ × ¾ in., and is screwed or nailed to B from below. A screw, G, is put into the side of E near the top. This screw has the underside of the head filed flat, and against this the screw, L, taps when the armature is attracted. *The arm, F,* which carries the armature, A, is 4½ × ½ × ½ in., and is pivoted by means of N, which passes through it and the uprights C. F must swing up and down freely. The hole for N, in this model, is 1¾ in. from the armature end.

[Pg 100]

207. *The armature* is fastened to F by a screw, S. A copper bur is put under the head of S to aid in keeping A from rocking sidewise.

Through F, and about half way between C and L, is put a screw, I, the lower end of which taps against the head of a screw, H, which is put into D. By unscrewing H a little, F will be raised, and A will be brought nearer the poles of M. *The rubber-band*, J, is placed over the head of I, and has tied to it a thread, O, which in turn is tied to a screw-eye, K. K screws into the end of B, and by turning it one way or the other, the tension, or pull, on J may be increased or diminished. There must be enough spring in J to pull A up after the current ceases; it must not pull so much that the magnet cannot draw A down hard enough to make a good click between L and G.

The Magnet, M, is explained in App. 89, and the construction of one bolt magnet is given in detail in App. 88. In this particular sounder the bolts are 2 in. long under the heads, thus bringing the tops of the bolt-heads about 2¼ in. above B. M is held to the base by a band of tin, T. The yoke may be screwed to B, as suggested in App. 90. This is the better plan.

208. Adjustment. You will find, although you make all of the parts with the dimensions given, that you will have to try, and change, and adjust before everything will work perfectly. A must not be allowed to touch the poles of M when it is pulled down, on account of the residual magnetism, which would keep it pulled down. Adjust this with F. The armature must not be pulled too far up from the poles of M by the tension in J; adjust this with I and H. If your battery is weak, the pull of J must be small, just enough to raise A.

The Battery. It is supposed, if you make an instrument like this, that you expect to use it for a line. In [Pg 101] that case make a regular gravity battery like the cell of App. 9. See Fig. 99 for line connections, and Fig. 98 for plan view of this sounder.

APPARATUS 124.

209. Telegraph Line; Connections. Fig. 99 shows the complete connections for our telegraph line, with two complete outfits. The capital letters are used on the right side, R, and small letters are on the left side, L. *The batteries*, B, b, are like App. 9. *The keys*, K, k, are like App. 119. The sounders, S, s, are like App. 121 or 122.

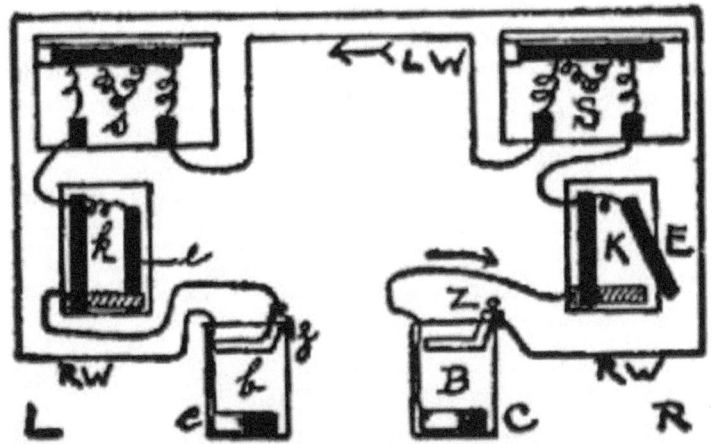

Fig. 99.

210. The two stations, *R* and *L*, may be near each other, or in different houses. The *return wire*, *R W*, passes from the copper of *b* to the zinc of *B*. This is *important*. If the cells are not joined properly, they will not work. It is better to have the cells together, on a short line, joined in series. The *line wire*, *L W*, and the return wire, *R W*, may be made of insulated copper wire for short lines in the house. Ordinary annunciator wire, No. 20, is good and cheap. The kind that is double cotton wrapped, waxed, and paraffined, has about 235 ft. to the pound. You should get at least 5 ft. for 1 cent. If your line stretches from one house to another you will find it better to use iron wire. Galvanized iron or steel wire No. 14 is good. This size [Pg 102] weighs about 100 lbs. to the mile. The return and line wires must not touch each other at any point; they must not touch any pipe or other piece of metal that will short circuit your batteries. It is best to use porcelain or glass insulators to support your wires if the line is long; but for short lines, where you use a return wire, you may support the wires upon poles or trees by means of loops made of strong cord or wire.

211. *Operation.* Suppose *R* (right) and *L* (left) have a line. By studying Fig. 99 you will see that *R*'s switch, *E*, is open while *e* is closed. The whole system, then, has but one place where the circuit is open. As soon as *R* presses his key, *K*, the circuit is closed, the

current from both cells rushes around through K, S, L W, s, k, b, R W, and B. This magnetizes the bolts of both S and s, and their armatures come down with a click upon the regulating-screws, where they remain as long as the current passes. As soon as R raises his key the armatures rise, making the up-click. R can, in this way, regulate the time between the two clicks. If he presses K down and lets it up quickly, the two clicks that his friend L hears from s are close together; this makes what is called a *dot*. If R holds K down longer, it makes a longer time between the clicks for L to hear, and this makes a *dash*. R, of course, hears his own sounder, which is making the dots and dashes also.

As soon as R has finished, he closes his switch, E. L then opens his switch and proceeds to answer. Both E and e should be left closed when you are through talking.

(Read § 194, 195, and study what is said in App. 9 about the gravity cell to be used on such a line.)

212. Telegraph Alphabet. The letters are represented by combinations of *dots, dashes* and *spaces*. A *dot* is made by pressing the key down, and raising it at once; [Pg 103] that is, the key is raised as soon as it strikes. This makes the letter E. The *dash* is made by pressing down the key, and allowing the current to pass about as long as it takes to make 3 dots; this makes the letter T. A long dash for L should take about as long as for 5 dots. *Spaces* occur in a letter and between words. To make a dash you hesitate while the lever of the key is down, to make a space, you hesitate while the key is up. H is made with 4 dots without hesitation or space. By putting a space between the dots the letter &, Y or Z is made according to the position of the space. Notice that letters containing dashes do not contain spaces. A space is really the opposite of a dash. The letters C, L, H, I, O, P, R, S, Y, Z, and & are made entirely of dots or of dots and spaces.

You should notice that several letters are the reverse of others; A is the reverse of N, B of V, D of U, C of R, Q of X, and Z of &. The student should study some book upon telegraphy, if he desires to become expert. Punctuation marks are left out of the alphabet here given, as boys will find very little use for them.

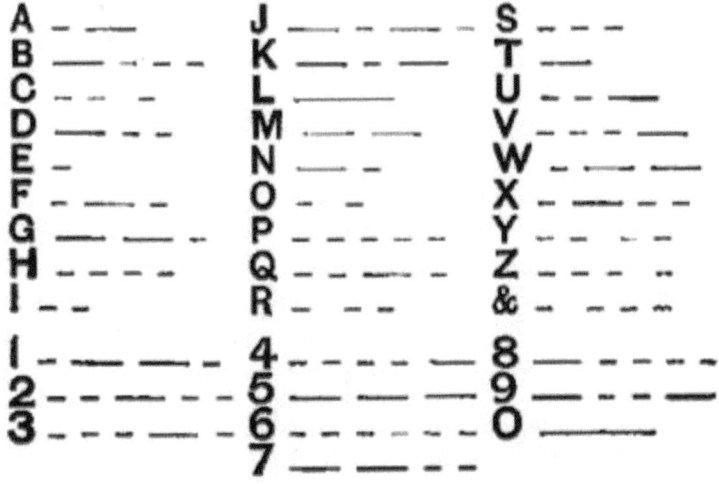

CHAPTER XV.

ELECTRIC BELLS AND BUZZERS.

APPARATUS 125.

213. Electric Buzzer. Fig. 100. A buzzer is, in construction, very similar to an electric bell; in fact, you will have a buzzer by removing the bell from any ordinary electric bell. They are used in places where the loud sound of a bell would be objectionable. As the buzzer is easier to make than a bell, we shall discuss it first.

214. The arrangement of the parts, (Fig. 100), is very much like that of the sounder of App. 121, Fig. 95. The armature is, in this case, a vibrating one and acts on the same principle as the automatic interrupter on App. 100, which you should study. (See § 148.) The general dimensions may be taken from App. 121. The base, B, in this case is about 1 in. wide. D also is made 1 in. wide. H is $1 \times 1 \times \frac{1}{2}$ in., and is nailed to A. Through its center is a hole for the regulating screw-eye, I. The end of I presses against F. The exact position of H will have to be determined after the magnets are in place. The arma-

ture, L, should be about ⅛ or 3/16 in. above the poles. They are not allowed to strike the poles, as a screw, E, regulates that. (See § 203). Y and Z are two binding-posts, like App. 46. To these are connected the battery wires. The strip of tin or copper, which forms Y, is cut like a letter T there being three holes in it, one near the end of each arm. The screw-eye, 2, and the screw, 3, are put through the horizontal part of the T, and the regulating-screw, I, passes through the hole in the vertical part [Pg 105] which springs up against I, thus forming an electrical connection between Y and I. The magnets are made and fastened as in App. 89.

215. *Connections.* The inside ends of the magnet coils, (§ 123), are fastened between a screw-head and a copper bur, S. One outside end goes to Z, and the other under the screw, G, which holds F to D.

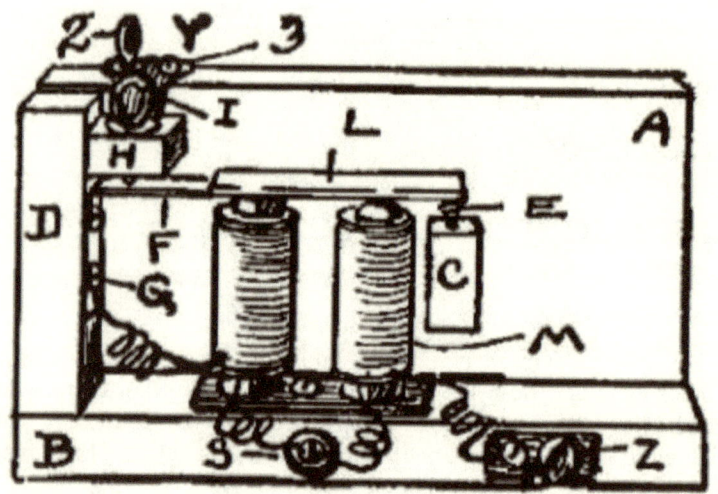

Fig. 100.

216. *Adjustment.* The part, F, and the screw, E, must be just high enough to keep L from striking the poles of M. If F is too weak, it will bend down to M. If F is too strong, it will take too much battery power to run it. In case there is not strength enough in F to quickly raise L when the current ceases to pass, arrange a screw-eye and rubber band as shown in Fig. 96. I should be *slowly* turned one way or the other, until it touches F just right to allow L to vibrate back and forth rapidly.

217. Operation. We shall suppose that you have all parts adjusted and the battery wires joined to Y and Z. If the current enters at Z, it will fly around through the coils, through G, F, up I, through the T-shaped tin and out at Y. The current was in L, but it could not get out at any other place than at Y. As soon as the bolts were magnetized, L was forcibly drawn down, pulling F away from I, [Pg 106] thus opening the circuit. As the bolts were no longer magnets, F sprang right back to I, the current passed long enough to re-magnetize the bolts. This operation was rapidly repeated.

218. Use. If you wish to use the buzzer simply to call some one occasionally, a dry battery or Leclanché cell is best. This apparatus is good to work a gravity cell when it needs regulating.

APPARATUS 126.

Fig. 101.

219. Electric Bell. Fig. 101. Before making this bell, carefully read the directions and explanations given for the electric buzzer, App. 125. The parts are very much alike in the two instruments, and most of the lettering of them has been made the same in the illustrations. If you look at Fig. 101 from the side, with the letters M and Q at the bottom, you will see that this bell is merely a modified form of App. 125.

The Base is $7 \times 5 \times \frac{1}{2}$ in. To the upper end of this [Pg 107] is nailed the cross piece, D. To D are fastened the binding-posts.

The Parts, $F, G, H, I, J, K, L, M, N, P, Q$, are the same as explained in App. 121 and 125.

The Magnet is fastened to the base by a tin strip, C, which is screwed down at both ends. By nailing a strip, like D, along the left side of the base, the magnet may be fastened to this. This strip would take the place of the base of App. 125.

The piece, F, of two thicknesses of tin, is made longer than it was in App. 125; in fact, it projects through L and forms the part N. To the lower end of N is fastened a large bullet. Hold the cutting-edge of a strong knife-blade upon the bullet, and with a few taps of a hammer drive the blade into it to make a gash.

Put the end of N into the cut, then hammer the bullet so that N will be pinched. If you have no bullet, cut a long strip of tin, about $\frac{3}{8}$ in. wide, and wind this about the end of N to serve as a ball.

The Bell, E, may be taken from an old alarm-clock. This is not screwed directly to the base, as it would not ring well. After you have the ball, O, properly fixed, hold E, so that O will strike it near its rim; then cut a piece of wood about $\frac{5}{8} \times \frac{5}{8}$, and long enough to put under E, to raise its rim to the right place. This piece must be screwed to the base from the underside, and on to its top is placed the screw which passes through the bell. In other words, E is mounted upon a rod which is fastened to the base.

The Adjustments are made as in App. 125. By bending N a little, O can be made to tap E properly.

The Battery for a bell that is to be used much should be an open circuit one, such as the Leclanché, or the ordinary dry batteries. It is cheaper to buy a dry battery than it [Pg 108] is to make one suitable for bells. *A* and *B* show wires that lead to the bell from the battery. One of the wires should be passed through a push-button.

APPARATUS 127.

220. Electric Bell. By arranging the buzzer of App. 125 with a bell, you can use the same for an electric bell. The part, *F*, should be made long enough to extend entirely through *L*, and project beyond *L* for about 2 in. To the end of this is fastened a large bullet, or a band of tin. (See App. 126.)

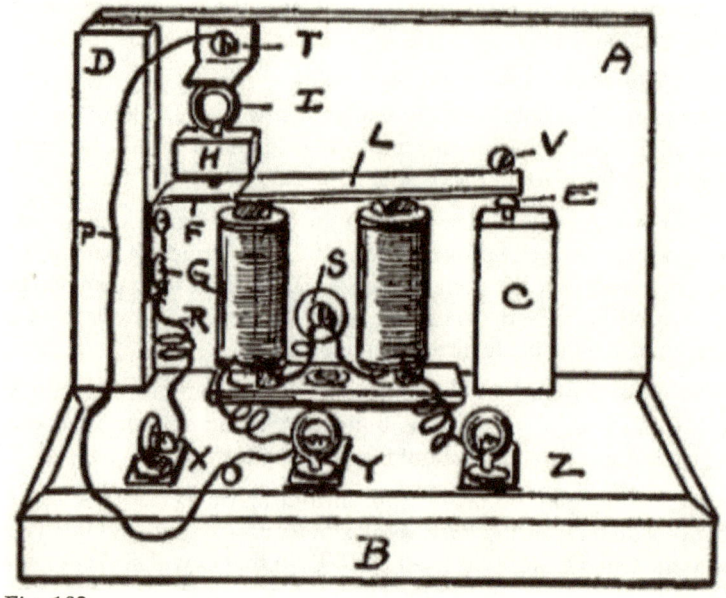

Fig. 102.

APPARATUS 128.

221. Combination Buzzer and Telegraph Sounder. Fig. 102. This apparatus is good for experimental purposes, where you do not wish to go to the trouble to make two separate pieces. For the dimensions and explanations see App. 121 and 125. There is but a slight change in App. 125 to make this.

222. *Connections.* The inside ends (§ 123) of the magnet wires are fastened together at *S*. The outside ends [Pg 109] are joined to the two binding-posts, Y and Z, made like App. 46. A wire, P, joins Y with the screw in T, which is a piece of stiff tin or copper, which presses down upon the top of I. In this way a connection may always be had between I and T. A wire, R, joins F electrically with X; it is held under the head of the screw, G. (See App. 125 about adjustments.)

223. *Operation.* When you wish to use the apparatus as a *buzzer*, join your battery wires to X and Z. If the current enters Z, it will pass through the magnet coils out to Y, through P, T, I, F, and R to X. If you use it as a *telegraph sounder*, join the battery wires to Y and Z. The current will then pass simply through the coils; it will not bother to go into P, F, etc., as it has no place it can escape. If used simply for experimental purposes almost any cell of sufficient strength will do. If for telegraph, use App. 9; if for buzzer, use an open circuit cell, as, for example, a dry cell.

[Pg 110]

CHAPTER XVI.

COMMUTATORS AND CURRENT REVERSERS.

224. *Commutators and Current Reversers* are useful in some experiments, as, for example, those with tangent galvanometers (App. 116, 117), in which readings are made with the current passing around the coil in one direction, and again made at once with the current reversed. The use of commutators on motors and dynamos should be understood. The reversers herein shown are, of course, not at all like those used on motors. Current reversers are used in connection with the needle-telegraph and many other instruments.

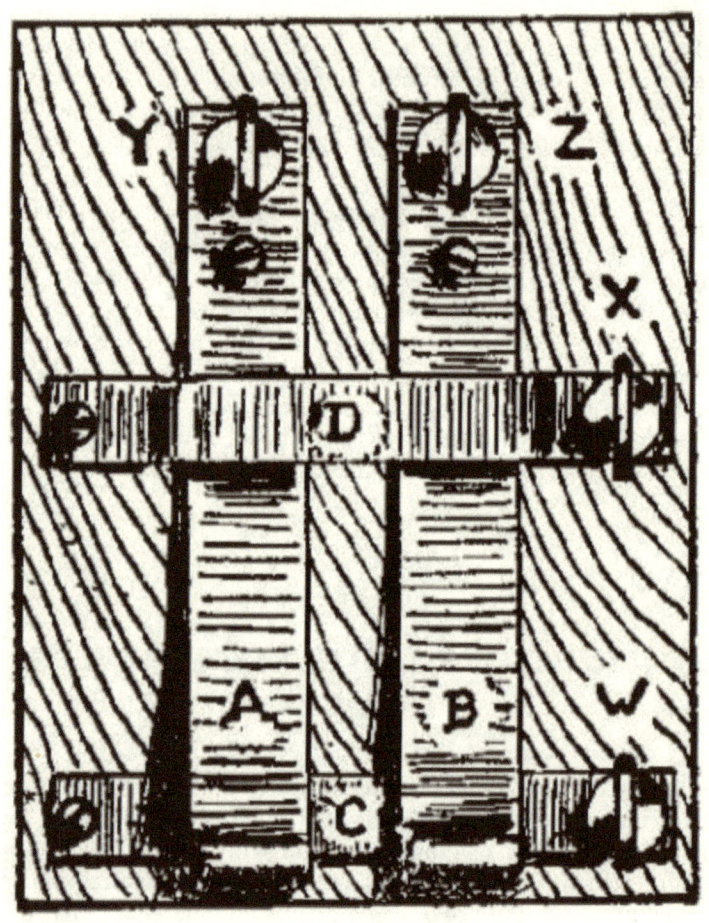

Fig. 103.

APPARATUS 129.

225. Current Reverser. Fig. 103. The *base* is 5 × 4 × ⅞ in. To this are fastened four *metal straps*, A, B, C, and D. These may be made of brass, aluminum, [Pg 111] or even of tin. If made of tin, use one thickness of metal for C and D, and two thicknesses for A and B. Each strap has two ⅛ in. holes punched in it, their positions being shown by the screw-heads and screw-eye binding-posts.

Construction. C is 3¾ × ½ in. Fasten this to the base first. At the left end is a small screw, while the right end is held down by the binding-post, W. The keys, A and B, should have quite a little spring to them. These are cut 5 × ¾ in. The front end of each is bent over a little (see the key App. 118, Fig. 92) so that they may be more easily grasped. The length after bending will be less than 5 in. The front ends should be raised from the base (Fig. 92) so that they will not touch C, unless pressed down. The ⅛ in. holes in the end of A are about ¾ in. apart, one being used for a screw to hold it to the base, and the other for the binding-post, Y. The strap, D, is 3¾ × ½ in. It is fastened at one end by a screw, and at the other end by X. D is bent about ¾ in. from each end, so that its middle part stands above the base about ¼ in. The straps, A and B, press up against D, unless they are held down with the hand.

226. *Connections.* W and X are joined to the poles of the battery to be used. Y and Z are joined to the apparatus in which the current must be passed in one direction, and then in the opposite direction. A tangent galvanometer, or a needle-telegraph instrument, for example, may be connected with Y and Z.

227. *Operation.* Suppose that the battery current enters at W. As long as both keys are raised, the current can go no farther. Now, imagine that we press A down solidly upon C, the current will pass along A, which does not now touch D, out through Y into the galvanometer, back to Z, into D, and to the battery again; that is, the current will enter the galvanometer from Y. Now, suppose that we let A spring up against D again, and press B down, the current still coming into W from the battery; the current will pass along B, out through Z, into the galvanometer, back to Y, through D, and back to the battery. It is evident, then, that the current can be made to pass out of Y or Z to the galvanometer at will by pressing down A or B.

APPARATUS 130.

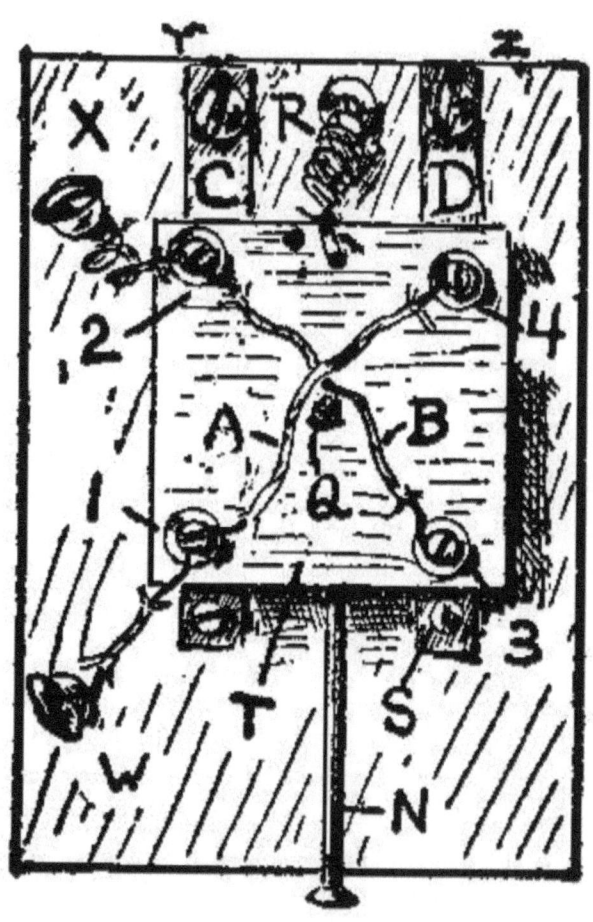

Fig. 104.

228. Current Reverser. Fig. 104. The wooden base is 7 × 5 × ⅞ in. To this are fastened two brass or tin straps, C and D, 5 × ½ in. They are fastened at the front ends by screws, S, while the binding-posts, Y and Z, hold the other ends solid. X and W are two screw-eye binding-posts (App. 45). The small square piece of wood, T, is 3 × 3 × ½ in. Through the corners of T, and in positions so that they will be directly over C and D, are put four screw binding-posts, 1, 2, 3, 4 (App. 41). The screws, however, pass entirely through T, and stick out about ¼ in. on the underside of it. The wire, A, connects W, 1 and 4, while the wire, B, connects X, 2 and 3. A and B must not touch each other where they cross on the top of T. N is a wire nail that serves as a handle. [Pg 113] If we were to place T, holding the four corner screws, upon the straps, C and D, it is evident that all the screws would touch the straps, if they were properly adjusted. We must fix things so that two only can touch the straps at a time. Put a screw, Q, through the center of T, from the bottom, so that it will stick out of the bottom more than the screws, 1, 2, etc. The screws, 2 and 4, will be lifted from C and D when the handle, N, is pressed down. By raising N, the top, T, can be made to rock up and down upon Q as a pivot. By lifting N far enough, 2 and 4 will be pressed against C and D, while 1 and 3 will be raised. A spring, R, is shown joined to T and to the base. This will hold the screws, 2 and 4, down upon C and D, unless N is pressed down.

229. Operation. We shall first suppose that the spring, R, is holding 2 and 4 in contact with C and D; 1 and 3 will, of course, be held up in the air. Imagine that we have a galvanometer connected with Y and Z. If the battery current enters at W, it will pass along A to 4, before it can find a chance to escape. It will pass through 4 into D, and into the galvanometer by way of Z, then back by way of Y, up 2, and out to the battery from X. If we now press the handle, N, down, the current will pass from W to 1, down 1 through C and Y to the galvanometer. It will return to the battery by way of Z, D, 3, B, and X. The current can then be rapidly reversed by raising and lowering N.

[Pg 114]

CHAPTER XVII.

RESISTANCE COILS.

APPARATUS 131.

230. Resistance Coils. Fig. 105. For experiments in resistance (See text-book), a set of *standard* resistances is necessary. There are many ways in which the resistances may be made; you can arrange them upon a long board, upon a rack, or wind the wires around spools. We generally speak of resistance *coils*. The Ohm is taken as the standard. If you use *copper* wire, you may take 9 ft. 9 in. of No. 30 insulated wire as *your* standard Ohm. You could, of course, take any other length of any size as *your* standard, but it will be best to make your coils with a certain number of Ohms resistance. If you have no No. 30 wire, you may use 39 ft. 1 in. of No. 24 insulated copper wire for 1 Ohm. (See wire tables in text-book.)

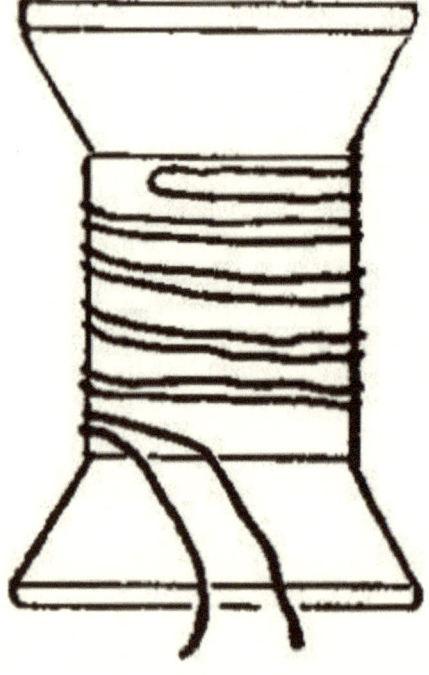

Fig. 105.

231. To avoid the magnetic effect (See resistance coils, in text-book), the wire should be measured off, then doubled, before winding it upon the spools. The wire may be held to the spool with paraffine. Fig. 105 shows how the [Pg 115] doubled wire looks on the spool, a few turns only being shown. Do not use any nails or other iron in connection with the coils proper.

232. By making 4 coils having, respectively, 1, 2, 2, and 5 Ohms resistance, you will be able to use any number of Ohms from 1 to 10. These will be very handy in connection with a "Wheatstone's bridge" for comparing resistances. (See text-book for experiments). The coils should be mounted upon a base with proper binding-posts, so that one or more coils can be used at a time. (See App. 132.) For the 2-Ohm coil use, of course, twice as much of the same kind of wire as for the 1-Ohm coil.

APPARATUS 132.

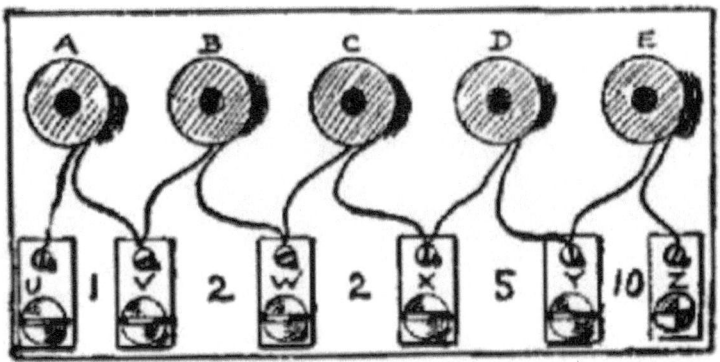

Fig. 106.

233. Resistance Coils. Fig. 106. The construction of one coil is given in App. 131. To have the set of coils so that they can be easily used, place the spools upon a base which, in the model, is 8½ × 4 × ⅞ in. The spools are 1¾ in. apart, center to center, and should be glued to the base. Fig. 106 is a plan of the apparatus. *U, V*, etc., are binding-posts like App. 46. The figures between them show how many Ohms resistance there are in the coil above. The coils *A, B, C, D,* and *E* are wound respectively for 1, 2, 2, 5 and 10 Ohms.

234. Connections. If you join a Wheatstone's bridge, [Pg 116] for example, with *U* and *V* (Fig. 106), the resistance added will be but 1 Ohm; if you join with *U* and *W*, the coils *A* and *B* will be in the circuit and make 3 Ohms resistance; if *V* and *X*, 4 Ohms; if *V* and *Y*, 9 Ohms; if *U* and *Z*, the whole, or 20 Ohms.

APPARATUS 133.

235. Resistance Coils. For use in some experiments in comparing the resistance, diameter, lengths, etc., of wires (See text-book), it is very handy to have coils made a certain number of meters long. (The meter is a French unit of measure and represents 39·3705 of our inches). German-silver wire has a much greater resistance than copper wire of the same size and length.

(*a*) Make a coil (See App. 131 for method) containing 1 meter of No. 30 German-silver wire.

(*b*) Make a coil with 2 meters No. 30 German-silver wire.

(*c*) Make one with 2 meters of No. 28 German-silver wire.

(*d*) Make one with 20 meters of No. 30 copper wire.

The above wire must be insulated if it is to be wound upon spools. Bare wire may be arranged on boards or racks so that the current may not be short circuited.

[Pg 117]

CHAPTER XVIII.

APPARATUS FOR STATIC ELECTRICITY.

236. Static or Frictional Electricity. There are many interesting and instructive experiments in this branch of electricity. All that can be done here is to explain a few pieces of simple apparatus to show the presence of static electricity, it being taken for granted that you know how to produce it, and that you have some book of simple experiments.

237. Electroscopes are instruments for showing the presence of static electricity.

APPARATUS 134.

238. Thread Electroscope. A piece of ordinary thread may be used for this purpose. Tie one end of it to the back of a chair or other support.

APPARATUS 135.

Fig. 107.

239. Pith-Ball Electroscope. Fig. 107. The pith from elder, cornstalk, milk-weed, etc., is very light and porous. When this is tied to the end of a silk thread, we get the pith-ball electroscope, so much talked about in nearly every text-book on physics. The upper end of the thread may be tied to any suitable support. Fig. 117 shows a book, lead pencil, and a small weight to hold the pencil steady. The thread is tied to one end of the pencil.

APPARATUS 136.

240. Support for Electroscopes, etc. Fig. 108. Glue or nail a spool, S, to a wooden base, B, measuring [Pg 118] about 4 × 5 in. Wrap some paper around a 7 in. length of ¼ in. dowel, D, to make it fit the hole in S. Wind one end of a wire, W, around the top end of D. To the outer end of W tie a *silk* thread, $S\ T$, on the lower end of which may be tied a piece of pith or material to serve as an electroscope.

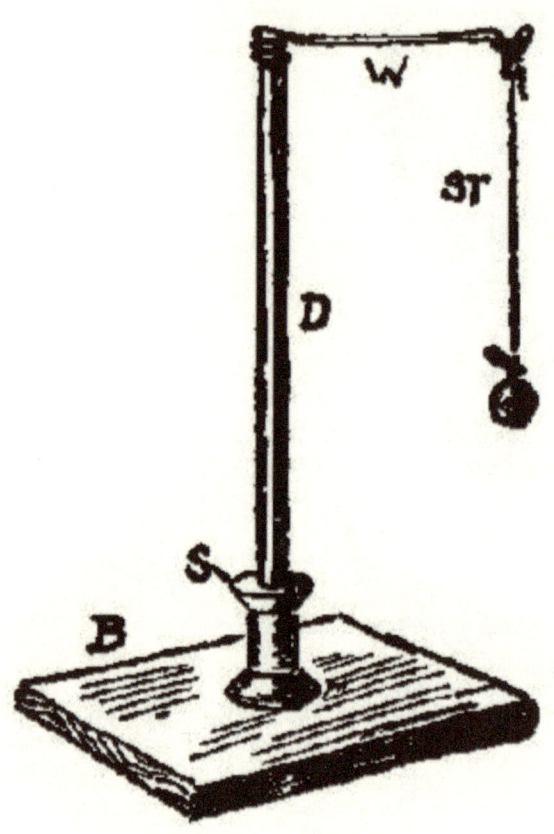

Fig. 108.

APPARATUS 137.

241. Carbon Electroscope. Carbon will be found to make a most excellent electroscope, as it is light and a good conductor of electricity. Light an ordinary match and let it burn until it is charred through and through. The black substance remaining is carbon. Tie a small piece of the carbon, about ¼ in. long, to one end of a silk thread, and support the thread as in Fig. 107 or 108.

APPARATUS 138.

242. Pivoted Electroscope. Fig. 109 and 110. Fold a piece of stiff paper double, then cut it into the shape shown. It should be about 3 in. long and 1 in. wide when opened out. A hole, B, about ½ in. in diameter should be cut in it while folded. A piece of paper, C, should be pasted to A, so that its top, where it is creased, [Pg 119] will be about ⅛ in. above the top of A. The support consists of a pin, E, stuck through a cork, D. Balance the paper on the pin, which passes up through the hole, B. An electrified body brought near this apparatus will make it whirl around very decidedly.

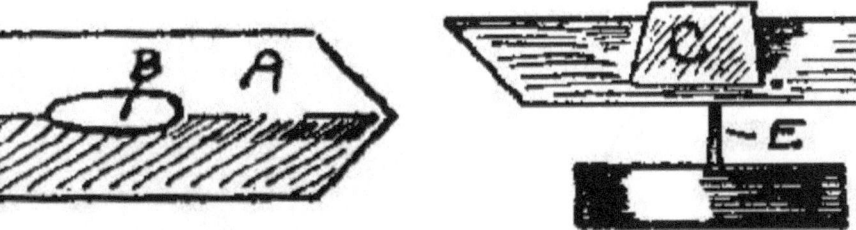

Fig. 109. Fig. 110.

APPARATUS 139.

243. Fancy Electroscope. Fig. 111. Fold a piece of stiff paper double, then cut out some fancy-shaped figure, as suggested, and draw the face, clothes, etc., to suit. This being folded through the center for cutting, it can be balanced upon a pin-point as explained in App. 138.

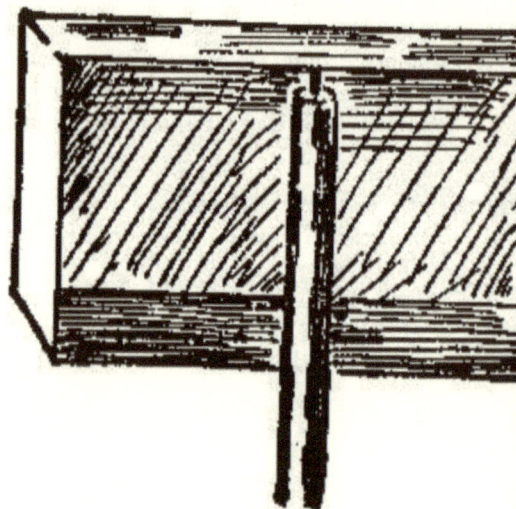

Fig. 112.

Fig. 111.

APPARATUS 140.

244. Box-Cover Electroscope. Fig. 112. A pasteboard box-cover, balanced upon a pin, makes a fairly good electroscope, although it is not nearly so sensitive as App. 138. The pin may be stuck in the upper end of the dowel, *D*, shown in Fig. 108.

[Pg 120]

APPARATUS 141.

245. Leaf Electroscope. Fig. 113. This is a very sensitive instrument, and can be used to tell the *kind* of static electricity on a body, as well as the mere presence of it. (See experiments in text-book.) The lamp chimney acts as a support for the leaves, *L*, and it protects them from currents of air. A tin box-cover, *C*, has a small hole punched through its center. Through this is pushed one end of a wire, *W*. This may be a hairpin, straightened. The upper end is bent over at right angles, after passing it through the hole. The lower end is bent as shown. On this horizontal part is fastened the leaf. These should be made of aluminum leaf, or of Dutch metal. The former will stand more rough handling than the latter. Goldleaf is used for very sensitive instruments. It is a little too delicate for unskilled hands.

Fig. 113.

Fig. 114.

246. To cut the aluminum leaf, place it between two pieces of paper, then cut paper and all into the desired shape. The piece should be about 3 in. long and 1 in. wide. Fold this across the middle, and stick it to the underside of the wire (Fig. 113). Saliva will make it adhere to the wire, if you have nothing better.

[Pg 121]

APPARATUS 142.

247. *To Show Where a Charge of Static Electricity Resides.* Fig. 114. This shows a tin baking-powder box placed upon a hot tumbler. A moist cotton thread is hung over the edge of the box. (See experiments in text-book.) The box will become charged by touching it with a charged body. The thread will show whether the charge resides upon the inside or upon the outside of the box.

APPARATUS 143.

Fig. 115.

248. *Support for Electrified Combs.* Fig. 115. In the study of static electricity, ordinary ebonite combs can be used to great advantage. A bent hairpin will serve as a cradle to support them. A silk thread may be tied to the wire, but a narrow silk ribbon is better than thread, as it will hold the comb steady.

[Pg 122]

CHAPTER XIX.

ELECTRIC MOTORS.

249. An Electric Motor is really a machine. If it be supplied with a proper current of electricity, its armature will revolve; and, if a pulley or wheel be fastened to the revolving shaft, a belt can be attached, and the motor made to do work. There are many kinds of motors, and many simple experiments which aid in understanding them. All that can be done here, however, is to show how to make simple motors. (See text-book for experiments.)

APPARATUS 144.

250. Electric Motor. Fig. 116, 117. Fig. 116 shows a plan or top view, and Fig. 117 shows a side view, with a part of the apparatus removed, for clearness.

The *base*, B, is 5 × 4 × ⅞ in. The *upright*, U, is 3½ × 1½ × ½ in., and is nailed or screwed to B. The *binding-posts*, X and Y are like App. 46. 4 is a screw binding-post.

251. The Field-Magnets, as the large electro-magnets on a motor are called, are made of 5/16 machine-bolts, 2½ in. long. The washers are 1½ in. apart inside. (See App. 88 for full directions.) The bolt cores are 2 in. apart, center to center. (See App. 89.) The tin yoke, D, is made like App. 71, and it is fastened to the base, like App. 90. The hole for the screw, however, is made a little to one side of the center, so that a dent can be made at the center for the bottom of the shaft, 8, to turn in. Make the dent with a center punch. The yoke is [Pg 123] fastened to B, so that one edge of it is 1½ in. from the back edge of B. (Fig. 116).

252. The Armature, A, is made of 6 or 8 thicknesses of tin, 2½ in. long and ¾ wide. (See App. 71.) In its center is punched or drilled a ¼ in. hole, so that it can be slipped onto the ¼ in. "sink-bolt," 8. If you have taps you can make the hole a little smaller than ¼ in., and thread it so that it will screw onto 8. A must be heavy enough to revolve a few times when once started. It is pinched between two nuts, 9 and 11, so that it just clears the poles when it turns. (See App. 145 for another form of armature.)

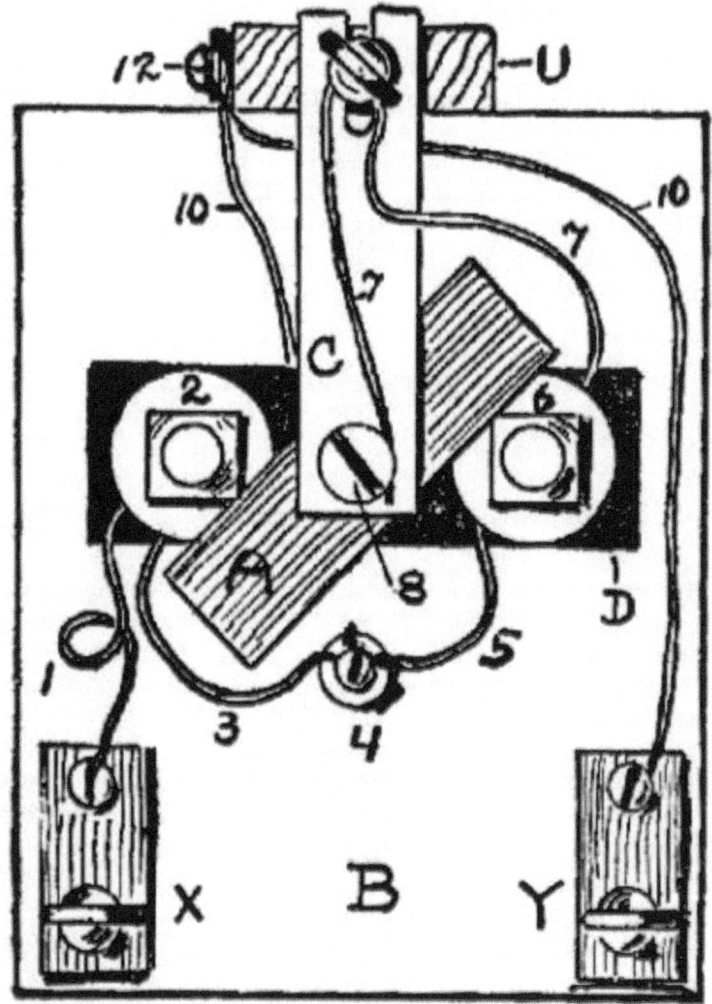

Fig. 116.

253. *The shaft or axle*, 8, is made of a "sink-bolt" that is 3 in. long and ¼ in. in diameter. These sink-bolts are threaded over their en-

tire length, and are furnished with two nuts, 9 and 11, Fig. 117. File or grind [Pg 124] the end of 8 to a point, so that it will turn easily in a dent made for it in the yoke, *D*, or in a dent made in another piece of tin fastened over the yoke. The shaft is held in a vertical position by the arm, *C*.

254. *The Arm, C,* is made of 2 or 3 thicknesses of tin. It is 3 × ¾ in.; it has in one end a hole for the shaft to revolve in easily, and in its other end a slot is cut. A screw-eye and bur are used to hold *C* to the upright, *U*. By this means the shaft can be moved and regulated as to position.

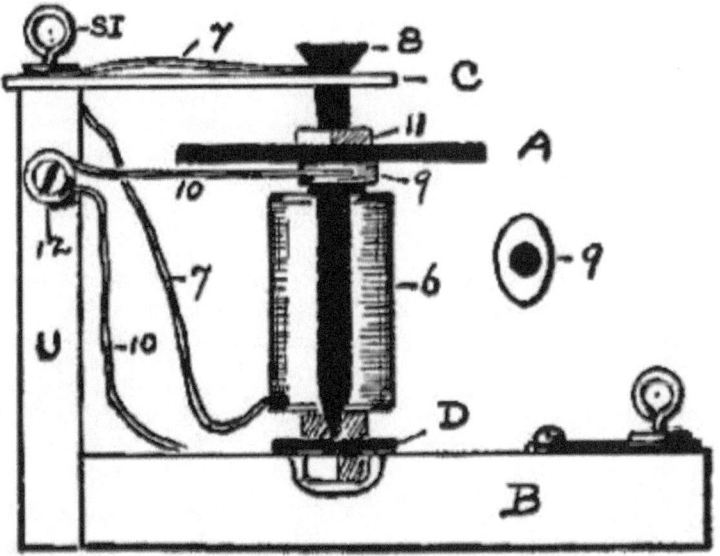

Fig. 117.

255. *The Commutator,* 9, (Fig. 117), is made of one of the nuts furnished with the shaft. Two of its corners are filed or ground off, so that it has the shape shown at the right, in Fig. 117. The copper wire, 10, rubs against 9, as the pointed part of it comes around. 10 is really a "brush," and carries the current into 9 at the right time.

256. *Connections.* Join the two inside ends (§ 123) of the coils to 4. The outside end of 2 is joined to *X*; the outside end, 7, of the other coil, 6, is carried up under or [Pg 125] around the screw-eye, *S I*, and

then its bare end reaches out and gently scrapes against the top of the shaft, 8. The wire, 10, leads from Y to the back of the base, where it is carried up to a screw, 12, which holds it to U. Its bare end reaches out to gently scrape against the commutator, 9, when it swings around. This wire, 10, should not press against 9 during the *entire* revolution.

257. *Adjustment.* Suppose the current enters at X. When the "brush," 10, presses against the commutator, 9, the current passes through X, 1, 2, 3, 4, 5, 6, 7, down 8 to 9, and out through 10 to Y. (The current, of course, goes down into D and into the bolt-cores also; but it can go no farther, if the coils are properly insulated, and A is not allowed to touch the cores. It is better to have the end of the shaft rest upon a piece of glass, having a slight depression made with a file, or in a dent made in tin which rests upon wood, the tin having no connection with D.) If 10 should continue to press against 9, the current would continue to pass, and A would be held firmly in place, directly over 2 and 6, and, of course, the shaft could not revolve. If, however, the brush leaves 9 (See plan of 9 at side of Fig. 117), just as A gets over the coils, or an instant before it gets there, the weight of A will carry it beyond the coils. No current should pass again, until A is at least at right angles to a line drawn through the center of the coils. If the current again passes, the ends of A will be attracted by the bolt-cores.

In other words, the current should pass a little less than one-half of the time, and this is divided into two parts. Suppose you start A with your finger; the current should be shut off automatically just before the center of A gets over the center of the bolt-cores. A makes ¼ of a revolution without current, and just after it gets beyond this, the current passes for nearly ¼ of a revolution, [Pg 126] which brings the ends over the poles again. The next ¼ of a turn it has no current, because the flat side of 9 is opposite the brush, 10, as during the first ¼. The last ¼ the current passes again. The exact position of the commutator will depend upon the way you arrange the brush. The positions of 9 and 10 can be found by trial, so that the circuit will be promptly opened and closed at the proper moment. Start the motor by turning the armature.

258. Batteries. The amount of power needed will depend upon how well you make the motor. One cell of App. 3 or 4 will run a well made one, but it is better to use 2 cells. Join the wires to X and Y.

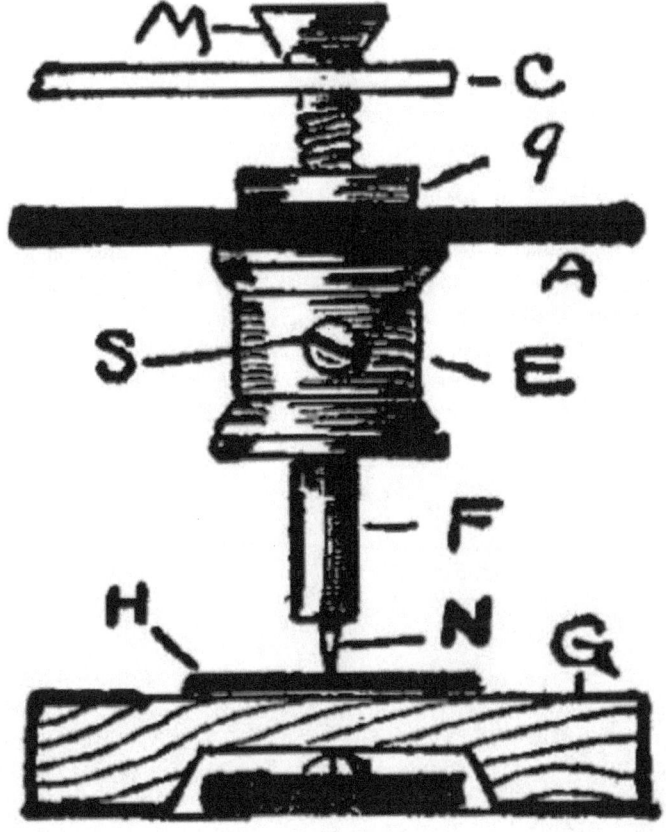

Fig. 118.

APPARATUS 145.

259. Armature for Motors. Fig. 118 shows another form of armature that may be used for small motors like App. 144; in fact, you may find that this form is easier to make than that of App. 144. *M* is a 5/16 machine screw, 1½ in. long, 9 being the nut furnished with it.

9 is filed as explained in § 255, and forms the commutator. *C* is the arm (§ 254). *A* is the armature [Pg 127] (§ 252). *A* is held firmly in place between the spool, *E*, and 9. *S* is a set-screw which passes through *E*, and holds the piece of ¼ in. dowel, *F*, in place. *N* is a needle-point fastened in the end of *F*. *N* revolves in a dent made in a piece of tin, *H*, which rests upon a wooden strip, *G*. *G* is cut away on its underside, so that it will straddle the yoke, *D*, Fig. 117; it is nailed to the base. This is given as a suggestion. By making *F* a little longer, *N* can turn in a dent made in the yoke, below *G*.

260. Adjustments. *M*, being 5/16 in. in diameter, will screw solidly into the hole in *E*. Place 9 upon it first, then *A*, and screw it about ½ way into *E*. 9 will serve as a lock-nut by turning it so that it will pinch *A* and hold it firmly against the top of *E*. *F* should reach half way into *E*. Put *N* in place after you have *H* and *G* arranged. You can then cut the upper end of *F* at such a place that it will bring *A* about ⅛ in. from the top of the magnet-cores. Paper wrapped around *F* will make a good fit in *E*. The current should enter *M* and leave 9, as fully explained in App. 144. (See § 257).

APPARATUS 146.

261. Electric Motor. Fig. 119, 120, 121, 122. Fig. 119 shows a front view, and Fig. 120 a side view of the whole motor. Fig. 121 shows the part that revolves, and includes the *shaft, armature* and *commutator*. Fig. 122 shows a section of the commutator. All the dimensions are taken from a model. You can modify the size to suit.

262. Wood-work. The *base* is 7 × 5 × ⅞ in. The *uprights*, *U*, are 3½ × 1 × ¾ in. They are screwed or nailed to the base from below, their 1-in. sides being towards you in Fig. 119. They are 4¼ in. apart, inside, in this model. The piece, *A*, is 2½ × ⅞ × ⅝ in., and is cut [Pg 128] away on the underside to straddle the yoke. Fig. 118 is a suggestion as to its shape. *A* is screwed or nailed to *B*.

263. Tin-work. The horizontal arm, *T*, is made of 3 thicknesses, and holds the shaft in a vertical position. *T* is 6¼ × ¾. In its ends are slots, and in its center is a hole so that the ¼ in. shaft can revolve easily, but not too loosely. The slots allow an adjustment, the screws, *S*, holding *T* to *U*. The shaft rests in a dent made in a piece of tin which is tacked to *A*. The yokes are elsewhere described.

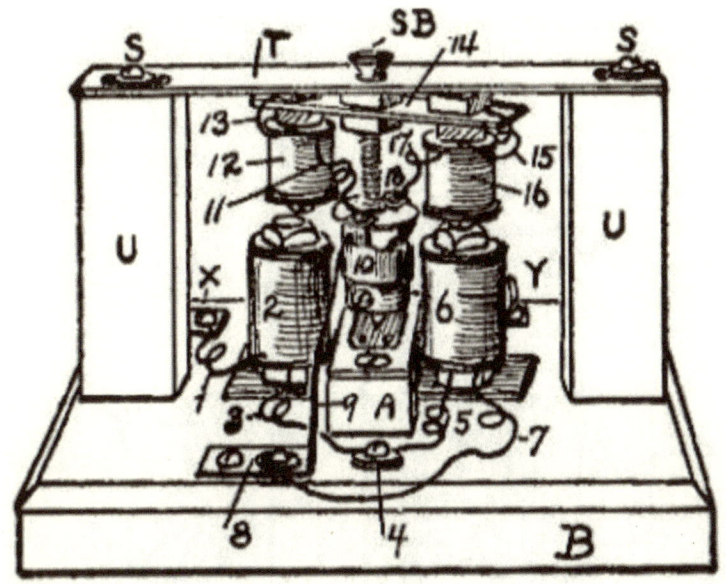

Fig. 119.

264. Field-Magnets. In this model they were made of 5/16 bolts, 2 in. long, placed 2 in. apart center to center. The washers are 1⅛ in. apart inside. (See App. 88 for full directions.) App. 89 and 71 should be studied. Except in size, they are made as in App. 144. They have 8 layers of No. 24 or 25 wire.

265. The Armature, Fig. 121, on this style of motor consists of a regular horseshoe electro-magnet, made in the same general way as the field-magnets. The electro-magnets, 12 and 16, are smaller, however, than the field-magnets. [Pg 129] The *cores* are ¼ in. stove-bolts, 1¼ in. long under the head. They are placed 2 in. apart, center to center. They are insulated and wound as fully explained in App. 88. These ¼ in. bolts require a change in your winder. (See App. 147 for this.) If you wish to use 5/16 bolts, you may use the same axle for your winder as before. The washers are ⅝ in. apart, inside. The cores are wound with 4 or 6 layers of No. 24 or 25 wire. This makes them about ¾ in. in diameter. They are held in a tin yoke, 14, made of 5 or 6 thicknesses of tin. 14 is 3 × ¾ in., and has 3 holes punched in it. The two outside holes are 2 in. apart. Through these pass the bolts, which are held firmly by the 2 nuts. The shaft, *S B*, is a sink-bolt, 3 in. long, and ¼ in. in diameter. (See § 253.) The inside ends (§ 123) of the coils should be firmly twisted together or held under the top nuts to make a good connection between them.

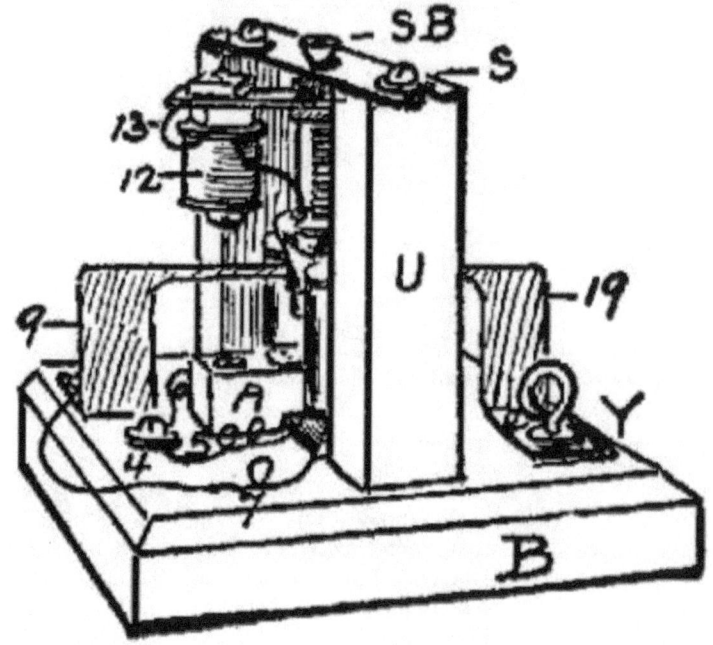

Fig. 120.

266. *The Commutator* is in two parts, which must be insulated from each other. The 2 sections are made out of thin tin or copper in the shape of an inverted T, as shown at 10, Fig. 121. The arms of the T are about ⅜ [Pg 130] in. wide, the horizontal ones reaching about half around the spool, E. The vertical arm reaches over the top of E, and is held down by a small screw, J. The sections, 10, must not touch the shaft. The outside wires (§ 123) of 12 and 16 are fastened under these screws, J, and they must not touch the shaft. Bend the tin sections so that they will be as nearly round as possible. The spool, E, has been sawed off so that it will go between the field-magnets. Wind paper around the shaft to make it fit solidly into E. S is a small screw that holds E in place, if the paper does not hold it tight enough.

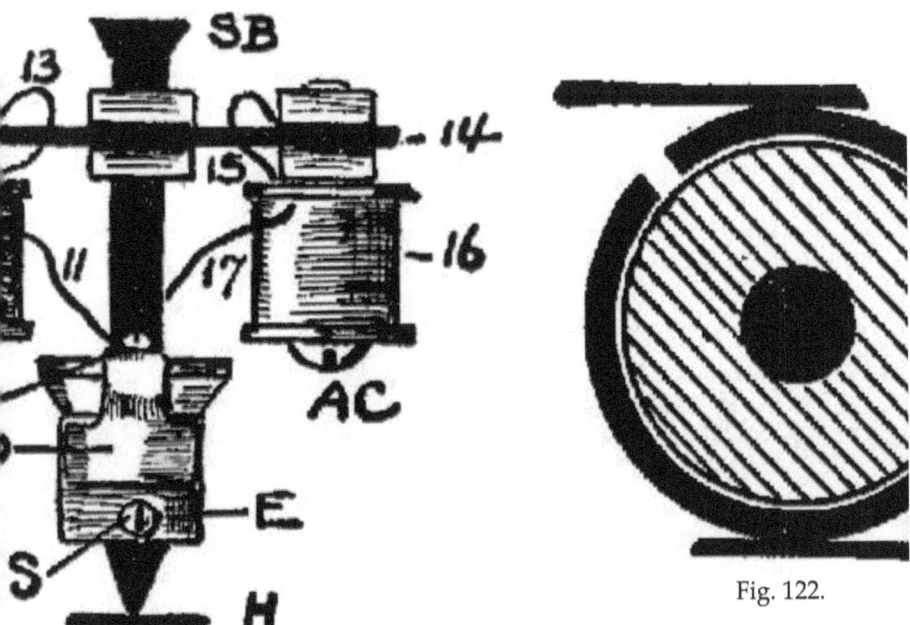

Fig. 121.

Fig. 122.

Fig. 122 shows a section of the spool and tin sections with the brushes pressing against them. The sections do not touch each other, and the brushes touch opposite sections. It is evident, then, that the current must pass through the coils 12 and 16 in order to get from one section of the commutator to the other, provided you have no short circuits through the shaft or elsewhere. The slots in the commutator must be directly under the center line of the yoke, 14, as seen in Fig. 121.

267. *The brushes*, 9 and 19, Fig. 120, are made of very thin tin or copper. They are cut to the shape shown, the narrow part being about ⅛ in. wide, and long enough to [Pg 131] reach at least to the center-line of the apparatus. The foot, or bottom part of the brushes, should be about 1¼ × ¾ in. These are used to fasten them to the base and to make connections. If you have no *thin* metal for brushes, use copper wires, and arrange them so that they will press gently against the commutator.

268. Connections. The inside ends (§ 123) of the field-magnets are held at 4. The outside end of coil 2 is joined to X, and that of coil 6 to 8, the foot of the brush which presses against 10. The section, 10, of the commutator is joined to 11, the outside end of coil 12, its inside end being fastened to the inside end of coil, 16, either by twisting them together, or by fastening them under the top nuts of the armature yoke, 14. The outside end of coil 16 is joined to the other commutator section, 18. The brush, 19, completes the circuit. In the foot of 19 is the binding-post, Y.

If the current enters at X, it will pass through 1, 2, 3, 4, 5, 6, 7, 8, 9, 10, 11, 12, 13, 14, 15, 16, 17, 18, 19, and out at Y, provided 10 and 18 are in contact with 9 and 19. Be careful not to have any short circuits. If, for example, the wire 7 touches 4, or if 3 touches 8, or if the wires 11 and 17 touch the shaft, your current will not pass where you expect, and you will have trouble.

269. Adjustment. The armature cores should just clear the poles of the field-magnets as they turn. This must be regulated by the thickness of A and the position of the nuts on the shaft, $S\ B$. The slots in the commutator must be under the center of the yoke, 14. The brushes, 9 and 19, must touch 10 and 18, but not so hard that they will stop the motor. Wire brushes are more easily adjusted than tin or sheet-copper ones. The tin arm, T, must hold the shaft properly. The point of the shaft must allow it to turn easily. The motor will turn [Pg 132] clockwise if the attachments are made as shown. Use 1 or 2 good bichromate cells, like App. 3 or 4.

270. Operation. The current will pass through the field-coils in the same direction, as long as the battery wires are not changed. The current is reversed in the armature-coils every time the brushes change from one section to the other of the commutator; that is, it flows in one direction during one-half of a revolution, and in the opposite direction during the other half. This reverses the poles of the armature-magnets every ½ revolution. (See text-book for full explanations and for simple experiments with electric motors.)

APPARATUS 147.

271. Attachment for Winder. In winding small electro-magnets for armature, etc., in which cores are used that are not 5/16 in. in diameter, your winder will have to be slightly changed. Its 5/16 stove-bolt will have to be removed, and a ¼ in. one put in instead. This may be done by making a handle for the ¼ in. bolt. To keep this from wobbling in the 5/16 hole, wind stiff paper around the bolt until it fits quite tightly. The whole winder is explained as App. 93.

[Pg 133]

CHAPTER XX.

ODDS AND ENDS.

APPARATUS 148.

272. *Graduated Circles.* Fig. 123. For compasses (App. 67), and for use in connection with tangent galvanometers (App. 116), a graduated circle is necessary. Fig. 123 is a reduced drawing from an original that is 4 in. in diameter. The long lines are 10 degrees apart, the smallest divisions shown being 5 degrees apart. Single degrees can be determined with considerable accuracy with the eye.

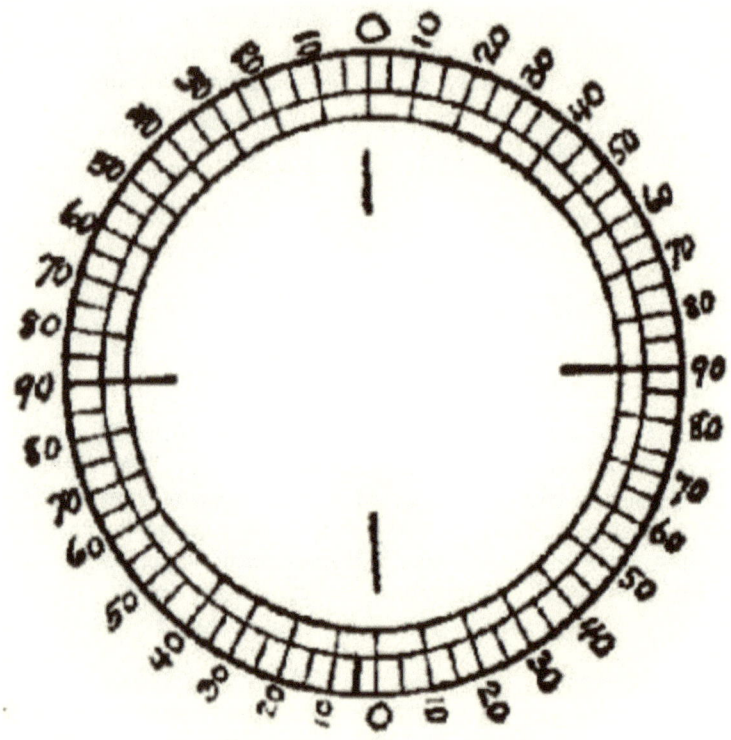

Fig. 123.

To divide the circle. Divide the circumference into 4 equal parts; these will be 90 degrees from each other, there being 360 degrees in every circle. Divide each quarter into nine equal parts with a pair of dividers; these will be for the long lines, 10 degrees apart. Divide each of these into two equal parts. If you are used to drawing, you can divide the circle still more, but 5-degree divisions will do.

[Pg 134]

APPARATUS 149.

273. *Adjustable Table.* Fig. 124. A table that can be raised or lowered is useful. The one shown at D, Fig. 124, is used for the galvanometer of App. 117. The dimensions are given in the figure. The upright piece, U, is fastened to D with brass screws, *not* with nails, as these would affect the needle. It is placed at one side of D so that the compass needle placed in the center of *D* will also be in the center of the wire coils when used in App. 117. The table is fastened in any position by a screw-eye, *S I*, which presses a copper washer, *W*, against *U*. *S I* works through a narrow slot, *S*, and screws into the back of the galvanometer. By making *S* longer, the table may be used for other laboratory purposes, if it is joined with some other form of standard.

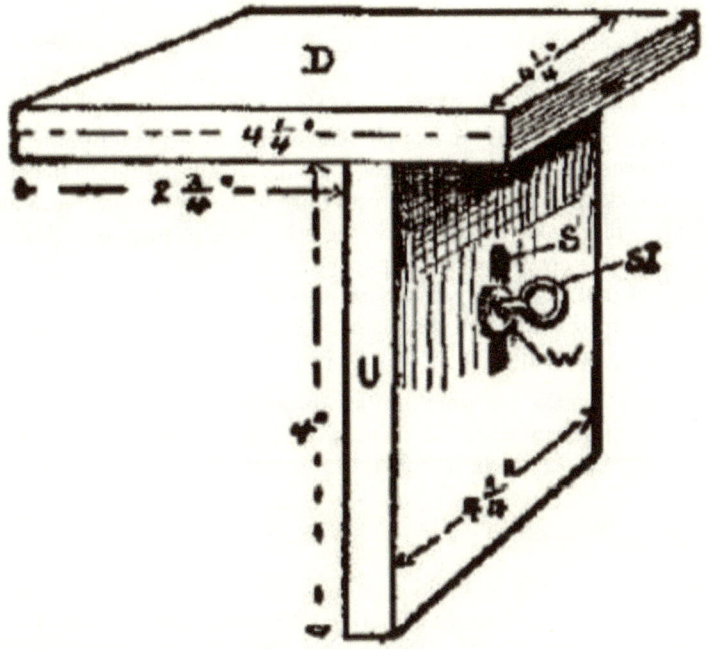

Fig. 124.

APPARATUS 150.

274. Glue Pot. If you have occasion to use glue, you can make a good glue pot out of 2 tin cans, one being placed inside the other. Put ¼ teacupful of glue in the inside can. If you have time, cover it with cold water, and let it soften. If you are in a hurry, cover it with hot water. Set this inside can into the other, in which you have boiling water. Do not let the water boil [Pg 135] over. The solder will not melt from ordinary tomato cans, if you keep water in them. Thin the glue with a little hot water until it drips from the brush in drops. Have the glue hot and fairly thin, and apply quickly. Hold the pieces of wood together by pressure until the glue hardens.

APPARATUS 151.

275. *Paraffine Paper and Cardboard* are extremely useful for insulating purposes. The paraffine used in candles will do, if you cannot get it in block form. While ordinary paper will do for simple apparatus to wind about coils, etc., you will find that paraffine paper can be handled very rapidly. To melt the paraffine you should use a double boiler, or one made of a shallow basin set in a pan of water. The water should be boiled. This will melt the paraffine in the basin. Strips of paper just passed through the melted paraffine will become soaked, and the paraffine will quickly harden in the air. Allow thick cardboard to soak for a minute or two, to drive out all the air. This makes excellent washers for electro-magnets. (See § 119.) To make one piece of this paper stick to another, merely pass a clean hot nail over the two where they lap. To hold coils of wire together, or to wooden bases, use a few drops of paraffine applied with a large hot nail.

276. Caution. Do not heat paraffine directly upon the fire or over a burner, unless you watch it constantly. It will burn if its temperature is raised too much. It is better to heat it with steam, as you do glue.

APPARATUS 152.

277. *Battery Jars.* For small cells, use glass tumblers. Ordinary glass fruit jars are good. Even earthen bowls may be used, and for large cells — if you [Pg 136] have nothing better — you can use small earthen crocks or jars.

278. Glass Bottles can be cut off so that they will make excellent jars. If you have thin bottles, you can cut them with strong cord. Tie one end of the cord, which should be 5 or 6 feet long, to a door knob or to a solid post. Tie the other end around your body. Make one complete turn of the cord around the bottle where you wish to cut it; draw the cord tight by stepping back, and with both hands draw the bottle back and forth vigorously many times, so that the cord will rub it hard and make it very hot. Do not let the cord move lengthwise upon the bottle. This will make a circle around the bottle that is very hot. Immediately plunge the bottle into cold water, the colder the better. Use ice-water, if you have it. If you produce heat enough, the bottle should crack all the way around very neatly. File off any sharp corners and edges with a wet file.

279. *A hot iron* can be used with success to cut off a bottle. File a deep groove first, hold the red-hot iron first on one side of file mark and then on the other to start the crack. You can lead the crack wherever you wish by keeping the iron about 1/8 in. ahead of it.

280. *A small gas-flame* will be much better than a hot iron, and you may easily use it, if you have glass tubing, rubber tubing, etc., in your shop. Draw out the glass so that the gas will burn in a fine needle-like flame about 1 in. long. Keep the point of the flame about 1/4 in. ahead of the crack. The glass tube should be held in a rubber tube connected with the gas pipe.

[Pg 137]

CHAPTER XXI.

TOOLS AND MATERIALS.

281. *Your Workshop.* If possible, keep all your work, tools and apparatus in one room, and lock the door when you leave.

The *work-bench* may be made of an old kitchen table, or of a strong, large box. The *tool chest* may be made of any clean box about the size of a soap box. *Shelves* can be made by setting soap or starch boxes on their sides, one above the other.

282. *The tools* needed are generally mentioned in the proper places, under the directions for construction. It is better to buy your tools as required, than to buy too many at once, some of which you may not need. If you have absolutely no tools, not even a saw or hammer, you will be obliged to buy or borrow, although a great deal can be done with a good knife. Do not be satisfied with rough-looking pieces of apparatus.

There are a few important tools needed for this work. While substitutes can be found for most of them, the boy who has access to a wood-working bench and tools will be able to do better and more rapid work than the boy who has no such tools.

283. List of tools. The following tools are needed, if rapid, accurate work is desired:

(1.) Lead pencil. (2.) A rule, divided into sixteenths for measuring. A straight foot rule will do, — cost one cent. (3.) Steel point for scratching lines on tin and copper. A stout needle-point is just the thing. (4.) An awl for making holes in wood; one that is a little less [Pg 138] than $\frac{1}{8}$ in. in diameter is best. (See App. 25.) (5.) A try-square with a 6 in. blade, so that you can mark out your apparatus with square corners. You can use a square-cornered box or piece of pasteboard, if you have no try-square. (6.) Chisels are very useful, but you can do wonders with a good sharp knife. (7.) Screw-driver. Do not use a good knife-blade for a screw-driver. (8.) A saw, one with teeth that are not too coarse is to be preferred. (9.) A plane is extremely useful to make your wood-work smooth and neat; but a great deal can be done with the sharp edges of broken glass, followed by a good rubbing with fine sand-paper. (10.) A brace and a set of bits may be needed in 2 or 3 cases, but nearly all of the holes can be made as in App. 25. (11.) Punches for sheet-tin, etc., will save much time. (See App. 26, 27.) For small holes in binding-posts, etc., use a flat-ended punch, $\frac{1}{8}$ in. in diameter. You should have one $\frac{1}{4}$ or 5/16 in. in diameter, if you make your yokes, armatures, etc., as in Chapter VIII. A blacksmith will help you out with this. (12.) A center-punch or sharp-pointed punch for making dents in metal. A sharp-pointed wire nail will do for tin and copper. (13.) Files for metal. (14.) Some sort of a vice or clamp. (See App. 79, 80.) (15.) Shears for cutting sheet-tin, etc. A pair of old shears will do. (16.) An anvil or piece of old iron that may be used to hammer on to flatten tin, etc. An old flat-iron makes a good anvil. (17.) Hammer.

The small hollow handle tool sets are very handy, and they contain small chisels, awls, screw-driver, etc. These sets cost from 50 cents up.

284. **Materials.** For *wood* you will find the sides and ends of clean soap or starch boxes about the right thickness; they are fairly smooth to begin with. For thin wood use cigar boxes. The pieces from old boxes [Pg 139] should be removed with care, and saved in one place, which may be called your lumber yard. All nails should be removed with a claw-hammer. Look out for nails when using a saw, plane or other edged tool. (See § 297.) The edges of bases, etc., may be bevelled as shown in Fig. 95. This is not necessary, but it adds greatly to the appearance.

285. Screw-Eyes. Brass screw-eyes, with copper burs, make excellent binding-posts. (App. 45, 46.) Those that are ⅜ in. in diameter inside the circle are about right. These are about 1¼ in. long in all, with a ½ in. thread.

286. Copper Burs, such as are used with rivets, are very handy. The size that is ½ in. in diameter, with a ⅛ in. hole, is good.

Fig. 125.

287. Copper Wire. This can be bought at an electrician's. The only trouble, however, in buying small quantities is that you may have to pay a large price in proportion. If you get it on ½ lb. spools you can handle it much better (see App. 23) than you can if you have it in a tangle. It is well to have ½ lb. of No. 24 or 25 for electro-magnets, current-detectors, etc., etc. ½ lb. of No. 30 will not be too much, if you make induction coils. If you handle your wire carefully, single cotton-covered will do. Double cotton-covered is better than single, but it costs more. Be careful not to injure the covering. (See below for splicing wire.) Look out for broken wire.

288. Splicing Wire. Fig. 125. Do not simply touch [Pg 140] two wires together and imagine that you have a good connection; a mere twist is not sufficient. Clean the ends of old wire thoroughly with a file or knife-blade, and join them as shown in Fig. 125.

289. Copper. Sheet-copper can be purchased at a tinsmith's or at a hardware store. Electricians usually have a thin variety of copper called brush copper, which makes good battery-plates, binding-posts, etc. You can cut this thin copper with an ordinary pair of shears.

290. Iron. For thin sheet-iron, nothing is better than sheet-tin. (See tin.) Hoop iron is thicker than tin, and makes good yokes, etc. In many cases, ordinary nails may be used where a magnetic substance is needed. Annealed iron wire is extremely soft. (See text-book for experiments with steel and iron.)

291. Steel. Old files, watch-springs, clock-springs, corset-steels, knitting-needles, harness-needles, hack-saw blades, sewing-needles, etc., are generally made of a good quality of steel.

292. Zinc, in the sheet form, can be bought at a hardware store. For a few cents you can get quite a large piece. Get the thick pieces for heavy battery-plates of an electrician. You do not need anything that is thicker than ⅛ in. The zinc rods are usually amalgamated.

293. Lead can be bought at a plumber's, tinsmith's, or hardware store. You may want some for a storage cell.

294. Nails. Wire nails are best for light work. Get an assortment from ½ in. long up to 1½ in.

295. Screws. It is better to use brass screws around electrical apparatus. For the small work, for binding-posts, etc., use ⅝ No. 5. Another handy size is No. 7, [Pg 141] from ¾ to 1¼ in. long. The round-headed screws are best, unless you want to countersink them.

296. Tin. This is really thin sheet-iron, covered with tin. Save up tomato-cans, cracker-boxes, condensed-milk cans, etc. The cracker-boxes are just as good as sheet-tin, as the pieces are large and clean. You can remove the solder from cans by heating them in the kitchen fire. Knock out the bottoms with a poker when the solder gets soft. Clean the tin with sand-paper.

297. Carbons. You can get carbon rods or plates at an electrician's. If you have arc electric lights in your city, you will be able to pick up carbons; these, however, generally have a coating of copper, which must be eaten off with dilute nitric acid. This is a bother. You will find it cheaper to buy the ½ in. rods that are 12 in. long, and uncoated.

298. Shellac. Your wood-work will be much improved by using shellac upon it after you have thoroughly sand-papered it. You can get it, all prepared, at a paint store. Wood-alcohol is used to thin it if it gets too thick. Keep it in a wide-mouth bottle. Paint it on quickly and evenly with a brush, and do not go over it again when it is partly dry. Wait until it is thoroughly hard before putting on a second coat. It should be fairly thin to spread well. Clean your brush in wood-alcohol before putting it away, and keep the shellac bottle tightly corked. A small tin can or a teacup is best to hold the shellac when using it.

www.ingramcontent.com/pod-product-compliance
Lightning Source LLC
Chambersburg PA
CBHW030031250526
45464CB00026B/1866